Digital Compensation for Analog Front-Ends

Digital Compensation for Analog Front-Ends

A New Approach to Wireless Transceiver Design

François Horlin
Université Libre de Bruxelles (ULB), Belgium

André Bourdoux
Inter-university Microelectronics Center (IMEC), Belgium

A John Wiley & Sons, Ltd, Publication

Copyright © 2008   John Wiley & Sons Ltd, The Atrium, Southern Gate, Chichester,
West Sussex PO19 8SQ, England

Telephone   (+44) 1243 779777

Email (for orders and customer service enquiries): cs-books@wiley.co.uk
Visit our Home Page on www.wiley.com

All Rights Reserved. No part of this publication may be reproduced, stored in a retrieval system or transmitted in any form or by any means, electronic, mechanical, photocopying, recording, scanning or otherwise, except under the terms of the Copyright, Designs and Patents Act 1988 or under the terms of a licence issued by the Copyright Licensing Agency Ltd, 90 Tottenham Court Road, London W1T 4LP, UK, without the permission in writing of the Publisher. Requests to the Publisher should be addressed to the Permissions Department, John Wiley & Sons Ltd, The Atrium, Southern Gate, Chichester, West Sussex PO19 8SQ, England, or emailed to permreq@wiley.co.uk, or faxed to (+44) 1243 770620.

Designations used by companies to distinguish their products are often claimed as trademarks. All brand names and product names used in this book are trade names, service marks, trademarks or registered trademarks of their respective owners. The Publisher is not associated with any product or vendor mentioned in this book.

All trademarks referred to in the text of this publication are the property of their respective owners.

This publication is designed to provide accurate and authoritative information in regard to the subject matter covered. It is sold on the understanding that the Publisher is not engaged in rendering professional services. If professional advice or other expert assistance is required, the services of a competent professional should be sought.

Other Wiley Editorial Offices

John Wiley & Sons Inc., 111 River Street, Hoboken, NJ 07030, USA

Jossey-Bass, 989 Market Street, San Francisco, CA 94103-1741, USA

Wiley-VCH Verlag GmbH, Boschstr. 12, D-69469 Weinheim, Germany

John Wiley & Sons Australia Ltd, 42 McDougall Street, Milton, Queensland 4064, Australia

John Wiley & Sons (Asia) Pte Ltd, 2 Clementi Loop #02-01, Jin Xing Distripark, Singapore 129809

John Wiley & Sons Canada Ltd, 6045 Freemont Blvd, Mississauga, ONT, L5R 4J3, Canada

Wiley also publishes its books in a variety of electronic formats. Some content that appears in print may not be available in electronic books.

Library of Congress Cataloging-in-Publication Data
Horlin, François.
   Digital Compensation for Analog Front-Ends : A New Approach to Wireless Transceiver Design / François Horlin, André Bourdoux
      p. cm.
   Includes bibliographical references and index.
   ISBN 978-0-470-51708-6 (cloth)
1. Radio–Transmitter-receivers–Design and construction. 2. Wireless communication systems. 3. Digital communications. I. Bourdoux, André. II. Title.
   TK6561.H63 2008
   621.3845–dc22

2008003743

British Library Cataloguing in Publication Data
A catalogue record for this book is available from the British Library

ISBN 978-0-470-51708-6 (HB)

Typeset by Sunrise Setting Ltd.

Contents

Preface   **ix**

1 Introduction   **1**
    1.1 Wireless transceiver functional description 1
    1.2 Evolution of the wireless transceiver design 3
         1.2.1 Independent design of analog front-end and digital transceiver 3
         1.2.2 Low cost analog front-end 3
         1.2.3 Higher system requirements 4
         1.2.4 Wish for software defined radios 4
         1.2.5 Technology scaling 4
    1.3 Contribution of the book 5
         1.3.1 Low-cost analog front-end 5
         1.3.2 Integrated system strategy 5
         1.3.3 Emerging wireless communication systems 5
    1.4 Organization 6

2 New Air Interfaces   **9**
    2.1 Orthogonal frequency-division multiplexing 9
    2.2 Single-carrier with frequency domain equalization 13
    2.3 Multi-input multi-output OFDM 15
         2.3.1 Space–time block coding 17
         2.3.2 Space-division multiplexing 18
         2.3.3 Space-division multiple access 21
    2.4 Code-division multiple access 23
         2.4.1 Direct-sequence code-division multiple access 23
         2.4.2 Multi-carrier code-division multiple access 24
         2.4.3 Cyclic-prefix code-division multiple access 25
    2.5 Frequency-division multiple access 28
         2.5.1 Orthogonal frequency-division multiple access 28
         2.5.2 Single-carrier frequency-division multiple access 30
    References 33

3 Real Life Front-Ends   **37**
    3.1 Front-End architectures 38
         3.1.1 Mathematical model of the ideal transmitter and receiver 38
         3.1.2 Classification of architectures 39

	3.1.3	Super-heterodyne architecture with analog quadrature	40
	3.1.4	Super-heterodyne architecture with digital quadrature	41
	3.1.5	Direct conversion architecture	43
	3.1.6	Low-IF architecture	44
	3.1.7	MIMO FE architecture	44
3.2	Constituent blocks and their non-idealities		44
	3.2.1	Amplifiers	44
	3.2.2	Mixers and local oscillators	45
	3.2.3	A/D and D/A converters	46
3.3	Individual non-idealities		47
	3.3.1	Nonlinear amplifiers	47
	3.3.2	Noise in amplifiers (AWGN)	53
	3.3.3	Carrier frequency offset	55
	3.3.4	Phase noise	56
	3.3.5	IQ imbalance	58
	3.3.6	DC offset	63
	3.3.7	Quantization noise and clipping	64
	3.3.8	Sampling clock offset	65
	3.3.9	Sampling jitter	67
References			69

4 Impact of the Non-Ideal Front-Ends on the System Performance   71

4.1	OFDM system in the presence of carrier frequency offset, sample clock offset and IQ imbalance		72
	4.1.1	Model of the non-idealities in the frequency domain	73
	4.1.2	Effect of carrier frequency offset	77
	4.1.3	Effect of sample clock offset	80
	4.1.4	Effect of IQ imbalance	82
	4.1.5	Combination of effects	84
	4.1.6	Extension to the frequency-dependent IQ imbalance	86
4.2	SC-FDE system in the presence of carrier frequency offset, sample clock offset and IQ imbalance		89
	4.2.1	Effect of carrier frequency offset	90
	4.2.2	Effect of sample clock offset	91
	4.2.3	Effect of IQ imbalance	93
	4.2.4	Combination of effects	95
4.3	Comparison of the sensitivity of OFDM and SC-FDE to CFO, SCO and IQ imbalance		95
4.4	OFDM and SC-FDE systems in the presence of phase noise		98
	4.4.1	System model	98
	4.4.2	Impact of the PN	99
	4.4.3	Numerical analysis	101
4.5	OFDM system in the presence of clipping, quantization and nonlinearity		103
	4.5.1	Clipping	103
	4.5.2	Quantization	106
	4.5.3	Clipping and quantization in frequency selective channels	107

		4.5.4 Spectral regrowth with clipping 108

- 4.5.4 Spectral regrowth with clipping 108
- 4.5.5 Power amplifier nonlinearity 110
- 4.5.6 Spectral regrowth with PA nonlinearity 111
- 4.6 SC-FDE system in the presence of clipping, quantization and nonlinearity . . 112
  - 4.6.1 Impact of quantization and clipping at the receiver 113
  - 4.6.2 Spectral regrowth with clipping at the transmitter 113
  - 4.6.3 Spectral regrowth with nonlinear PA 116
- 4.7 MIMO systems 117
  - 4.7.1 Impact of CFO and SCO on MIMO-OFDM 117
  - 4.7.2 Sensitivity of STBC and MRC to CFO 120
  - 4.7.3 Antenna mismatch and the reciprocity assumption 122
- 4.8 Multi-user systems 126
  - 4.8.1 MC-CDMA versus OFDMA 128
  - 4.8.2 User dependent non-idealities 131
- References 133

5 Generic OFDM System   135
- 5.1 Definition of the generic OFDM system 135
  - 5.1.1 Frame description 136
  - 5.1.2 OFDM receiver description 138
- 5.2 Burst detection 140
  - 5.2.1 Energy-based detection 141
  - 5.2.2 Auto-correlation-based detection 143
- 5.3 AGC setting (amplitude estimation) 144
- 5.4 Coarse timing estimation 146
- 5.5 Coarse CFO estimation 149
- 5.6 Fine timing estimation 150
- 5.7 Fine CFO estimation 151
- 5.8 Complexity of auto- and cross-correlation 152
- 5.9 Joint CFO and IQ imbalance acquisition 153
  - 5.9.1 System model 154
  - 5.9.2 Likelihood function and its second-order approximation 157
  - 5.9.3 ML estimate of the CFO in the absence of IQ imbalance 158
  - 5.9.4 EM algorithm for the joint CFO and IQ imbalance estimation . . . 159
  - 5.9.5 Implementation 161
  - 5.9.6 Performance and complexity analysis 163
  - 5.9.7 Compensation of the CFO and IQ imbalance 168
- 5.10 Joint channel and frequency-dependent IQ imbalance estimation 168
  - 5.10.1 System model 169
  - 5.10.2 ML channel estimation 170
  - 5.10.3 Performance and complexity analysis 171
  - 5.10.4 Frequency-dependent IQ imbalance compensation 172
- 5.11 Tracking loops for phase noise and residual CFO/SCO 174
  - 5.11.1 Estimation of residual CFO/SCO 174
  - 5.11.2 CFO tracking loop 175
  - 5.11.3 SCO tracking loop 179
- References 182

6 Emerging Wireless Communication Systems   185
   6.1   IEEE 802.11n 186
         6.1.1   Context 186
         6.1.2   System description 188
         6.1.3   Main challenges and usual solutions 194
         6.1.4   Compensation of non-reciprocity 203
   6.2   3GPP Long-term evolution 205
         6.2.1   Context 205
         6.2.2   System description 207
         6.2.3   Main challenges and usual solutions 210
         6.2.4   Advanced channel tracking 212
   References 223

Appendices   225

A   MMSE Linear Detector   227
   References 228

B   ML Channel Estimator   229
   References 230

C   Matlab Models of Non-Idealities   231
   C.1   Receiver non-idealities 231
         C.1.1   Global RX non-idealities 231
         C.1.2   Receiver noise 233
         C.1.3   Carrier frequency offset 234
         C.1.4   Phase noise 234
         C.1.5   AGC 237
         C.1.6   Receive IQ imbalance 237
         C.1.7   Sampling clock offset 238
         C.1.8   Clipping and quantization 241
   C.2   Transmitter non-idealities 242
         C.2.1   Global TX non-idealities 242
         C.2.2   Clipping and quantization 243
         C.2.3   Transmit IQ imbalance 244
         C.2.4   Phase noise 245
         C.2.5   Carrier frequency offset 245
         C.2.6   Sampling clock offset 245
         C.2.7   Nonlinear power amplifier 245

D   Mathematical Conventions   247

E   Abbreviations   249

Index   253

Preface

The joint design of the analog front-end and of the digital baseband algorithms has become an important field of research for a few years because it enables the wireless system and chip designers to better trade the communication performance with the production cost. Unfortunately the existing designs apply this approach rather opportunistically to solve a well-determined problem. There is clearly a lack of a global approach.

The aim of this book is to propose a systematic approach to design a digital communication system. In particular, we will present how our methodology can be applied to the emerging wireless communication systems. As such, this book will be a valuable reference for wireless system architects and chip designers.

More generally, our book intends to be cross-disciplinary and to cover in detail the digital compensation of many non-idealities, for a broad class of broadband emerging standards and with a system approach in the design of the receiver algorithms. In particular, system strategies for joint estimation of synchronization and front-end non-ideality parameters will be emphasized. This approach is actually linked with the in-depth expertise that has been developed in the wireless research group of IMEC where the authors have spent many years and have been involved in projects covering the main broadband wireless standards.

The organization of the book is also very important to bring the reader up-to-date with the main topic and to assist him/her in gradually absorbing the important and vast material. We cover in the first chapter a detailed introduction of the emerging wireless standards, which is essential in understanding the rest of the book, followed in the second chapter by a detailed description of the front-end non-idealities. From this point, the reader is well equipped to understand what happens when the topics described in the first two chapters are merged, which is the goal of the third chapter. The last two chapters continue with an in-depth coverage of the estimation and compensation algorithms, first for a generic system to understand the methodology and details of the system approach, then for two main emerging standards to be more pragmatic and fully in line with the real world.

1

Introduction

1.1 Wireless transceiver functional description

Emerging wireless communication systems are carefully designed to optimize at the same time the offered user capacity, the average spectral efficiency and the cell coverage. A set of complementary functions are successively implemented at the transmitter and at the receiver to support the communication while still respecting the power consumption and spectral occupancy constraints. Figure 1.1 gives a functional description of a typical wireless transceiver. Part of the functions are implemented on a digital processor (block (A) at the transmitter and block (D) at the receiver). The other part of the functions are implemented in the analog front-end (block (B) at the transmitter and block (C) at the receiver). In the following paragraphs, we give a synthetic description of the main transceiver functional blocks.

The *channel coder* adds structured redundancy to the bit stream at the transmitter that can be used at the receiver by the *channel decoder* to detect and ultimately correct the bit detection errors generated by various sources of signal distortion in the system. The *interleaver/deinterleaver* pair makes sure that the errors happen at random locations in the bit stream (bursts of errors are avoided).

At the transmitter, the *constellation mapper* converts the stream of coded bits into a stream of complex symbols. The necessary physical bandwidth is directly proportional to the symbol rate, so that high constellation orders are often foreseen to improve the system spectral efficiency. At the receiver, the stream of estimated symbols is converted back by the *constellation demapper* into a stream of estimated bits.

Most of the emerging communication systems rely on the orthogonal frequency division multiplexing (OFDM) technology (or on a derivative technology) to cope better with the channel time dispersion. The stream of complex symbols is organized in blocks of symbols (*serial-to-parallel converter*), that are possibly processed with the *linear pre-coder* and mapped with the *carrier mapper* onto a set of equally spaced sub-carriers in the frequency domain. The *inverse fast Fourier transform* (*IFFT*) transforms the blocks to the time domain and the *cyclic prefix* (*CP*) *adder* repeats part of the resulting blocks to make the transmitted signal appear periodic. Finally the time domain blocks of complex samples are converted to

Digital Compensation for Analog Front-Ends François Horlin and André Bourdoux
© 2008 John Wiley & Sons, Ltd

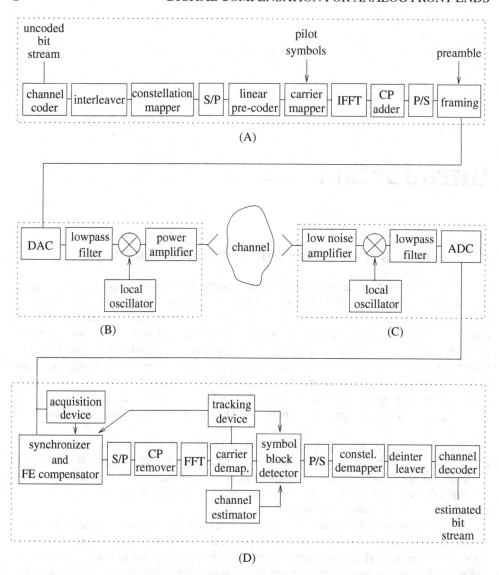

Figure 1.1 Wireless system block diagram: (A) transmit digital transceiver; (B) transmit analog front-end; (C) receive analog front-end; (D) receive digital transceiver.

a sequence of complex samples (*parallel-to-serial converter*). At the receiver, the counterpart of each operation defined at the transmitter is successively performed. In particular, the fast Fourier transform (*FFT*) brings back the signal to the frequency domain, where the time dispersive channel is seen as a multiplicative channel that can be compensated at a low complexity by the *symbol block detector*.

INTRODUCTION

The functions described until now are all digital and therefore take a digital sequence as input/output. However, the signal sent over-the-air at the antenna can only be a time continuous signal. The role of the transmit analog front-end is to convert the sequence of complex samples into a time continuous signal sent at a pre-defined transmit power level within the allocated frequency range. The sequence of complex samples is first converted with the *digital-to-analog converter* into an analog signal, filtered afterwards with the *lowpass filter* to limit the signal to the desired frequency range, translated to the carrier frequency generated by the *local oscillator* with the *mixer*, and finally amplified to a sufficient power level with the *power amplifier*. The continuous signal received at the receiver antenna is first amplified with a *low noise amplifier* that adapts the signal power to the required receiver power input range, transported back to the baseband domain with the *mixer*, filtered with a *lowpass filter* to avoid aliasing when the signal is sampled in the *analog-to-digital converter*.

Before the communication can effectively happen, the receiver should be synchronized to the transmitter. The role of the *acquisition device* is to estimate the signal time reference and carrier frequency, together with some front-end parameters, based on a known *preamble* transmitted in front of the data burst. The error measured according to the transmitter reference is compensated during the *synchronization*. Because the acquisition is not perfect (noise corruption...), the remaining errors have to be continuously estimated in the *tracking device*. Known *pilot symbols* are often interleaved in the data symbols to support the tracking.

In this first description of the transceiver, we have clearly separated the digital and the analog functions. It should be mentioned, however, that some functions can be implemented either on the digital processor or in the analog front-end, or even shared between the two.

1.2 Evolution of the wireless transceiver design

Even if the digital transceiver and analog front-end were initially designed separately, there is today a clear trend to integrate the two designs in order to better trade the cost and power consumption of the final products. This section presents how the evolution of the wireless systems has triggered the development of a new field of expertise in the design of mixed-signal solutions.

1.2.1 Independent design of analog front-end and digital transceiver

Originally, the analog front-end and the digital baseband transceiver were designed separately. The role of the analog front-end was to down-convert the signal from the bandpass domain to the baseband domain. The role of the digital baseband transceiver was to demodulate the digital transmitted signals taking the effects of the propagation channel into account. Different competencies were needed for the design of the analog front-end and for the design of the digital baseband transceiver. Each of the two blocks was designed by assuming that the other block works ideally.

1.2.2 Low cost analog front-end

The desire to build lower cost analog front-ends has triggered the interest for a new domain of research. Indeed, by observing the impact of the analog front-end non-idealities on the received signal, it has been shown that some of them could be compensated digitally. As an example, the correlated part of the carrier frequency jitter causes a phase rotation on the received samples that could be estimated and removed from the received signal. Therefore, it is interesting to tolerate a certain level of non-ideality in the analog front-end, making it much cheaper, that can be compensated afterwards in the digital domain at a reasonable complexity. There exists a point of convergence between the analog and digital experts: they have to define together the specifications on the analog front-end (tolerable value of each non-ideality) making the optimal compromise between the cost of the analog components and the digital complexity of the compensation algorithms.

1.2.3 Higher system requirements

The emerging wireless communication systems employ higher bandwidths, higher constellation orders and multiple antennas to meet the new user needs. As a result, the system becomes more sensitive to the front-end non-idealities. For example, the difference between the lowpass filters on the I and Q branches becomes more pronounced, making the IQ imbalance frequency-dependent. Another example comes from the amplitude and phase mismatch between the different antenna analog front-ends. Therefore, the front-end compensation techniques should be designed carefully in the new communication systems.

1.2.4 Wish for software defined radios

Future wireless systems of the fourth generation (4G) will integrate different existing and evolving wireless access systems that complement each other for different application areas and communication environments. To enable seamless and transparent inter-working between these different wireless access systems, or communication modes, multi-mode terminals are needed that support existing as well as newly emerging air interface standards, thereby offering a trade-off between the data rate, the range and the mobility. A first approach to build a multi-standard terminal is to duplicate the analog front-end. It is therefore clear that the cost of each analog front-end becomes critical and that the techniques to compensate the front-end non-idealities are required to limit the cost of the overall system. A second approach to build a multi-standard transceiver is to construct a flexible analog front-end. By tuning a set of parameters, the analog front-end will support a given communication mode. Once again, flexible analog front-ends are very costly and all efforts should be made to reduce their cost. Therefore, the digital compensation techniques of the non-ideal analog front-end are also needed.

1.2.5 Technology scaling

Much effort is spent today in the minimization of the production cost and power consumption of the new hardware platforms. In line with this trend, the dimension of the transistors is ever decreasing. Unfortunately, the scaling of the technology also brings new challenges in terms of analog front-end non-idealities. As the supply voltage decreases when the dimension of the transistors is reduced, the dynamic range is limited and the level of non-linearities increases. The behavior of the components is also less predictable due to their increased variability. On the other hand, the computation power of the digital platforms is improved and more complex front-end compensation algorithms are enabled.

INTRODUCTION 5

1.3 Contribution of the book

The joint design of the analog front-end and of the digital baseband algorithms becomes more and more popular in the literature. Unfortunately the existing designs apply this approach rather opportunistically to solve a well-determined problem. There is clearly a lack of a global approach. The aim of this book is to handle the problem globally by proposing a systematic approach to design a digital communication system. This section explains the three major objectives targeted in this book.

1.3.1 Low-cost analog front-end

The primary goal of the book is to describe how the non-idealities introduced by the analog front-end elements can be compensated digitally. As an example, the phase noise (or carrier jitter), caused by the inaccuracy of the local oscillators, and the IQ imbalance, caused by the difference between the analog elements on the I and Q branches when analog frequency down-conversion is applied (zero-IF architecture), will be considered. By doing so, the specifications on the analog components can be relaxed and the overall analog front-end can be made at a lower cost.

1.3.2 Integrated system strategy

The book presents the joint strategy for the front-end compensation, channel estimation and synchronization. The different non-idealities are estimated, often based on known sequences of transmitted symbols, and compensated afterwards. The non-idealities can be first estimated based on a preamble sent at the beginning of a physical burst, used also for the mobile terminal coarse synchronization (time and frequency) and for the channel estimation. Another possibility is to estimate them by means of pilots sent simultaneously with the information symbols, used also for the tracking of the synchronization errors and channel changes over the time. It is therefore important to integrate the estimation and the compensation of the non-idealities with the synchronization and channel estimation algorithms. Because each system has its own requirements, the strategy for joint front-end compensation, channel estimation and synchronization is specific to the system under interest.

1.3.3 Emerging wireless communication systems

In this book, we will study how the transceivers can be designed for emerging wireless communication systems to enable the compensation of the analog front-end. The methodology will be applied to the emerging wireless local area network (WLAN) and cellular communication systems:

- As for the WLAN communication systems, the IEEE 802.11n will be studied. It is the multiple antenna extension of the existing WLAN IEEE 802.11a/g based on OFDM.

- As for the cellular communication systems, the 3GPP LTE will be investigated. It targets much higher data rates than the third-generation cellular communication systems under much higher mobility conditions. A hybrid air interface has been selected: the downlink will be based on OFDMA and the uplink will be based on SC-FDMA to lower the constraints on the mobile terminal power amplifier.

1.4 Organization

We will first present the main air interfaces that are emerging in the wireless communication systems and summarize their properties. Second, a model of a non-ideal analog front-end is introduced, based on which the impact of the non-idealities on the performance of the air interfaces will be evaluated. The methodology to estimate and compensate each non-ideality will be explained by considering a generic communication system as a reference for the sake of clarity. Afterwards, the methodology will be extended to the emerging IEEE 802.11n and 3GPP LTE communication systems.

- The most promising air interfaces for the emerging wireless communications systems will be introduced in Chapter 2. For each air interface, a simplified block diagram of the system will be provided, based on which a mathematical input/output relationship will be built. The major principles will be explained (frequency domain multi-path channel equalization, spatial diversity with multiple antennas...). Finally, the air interface will be integrated in the most adapted communication system (WLAN or cellular) and reference performance curves will be provided.

- In Chapter 3, typical super-heterodyne and direct down-conversion front-end architectures will be introduced, and extended to multiple antennas. Based on that, the constituent blocks will be investigated and their non-ideal behavior will be characterized. Finally, a mathematical description of the different non-idealities will be provided, based on which a front-end model can be built.

- The impact of the non-idealities on the different air interfaces will be studied in Chapter 4. For each non-ideality, an analytical description will be provided (impact on the constellation, error distribution) and a numerical analysis will be performed. We will look into the individual impact of the carrier frequency offset, the phase noise, the sample clock offset, the clock frequency jitter, the IQ imbalance, the clipping and quantization and the non-linearities. Afterwards, we will investigate the effect of the non-idealities when they are introduced jointly in the system. In the case of multi-dimensional systems (multiple antennas and multiple users), we will evaluate the inter-stream interference generated by the different effects. The impact of the mismatch between the different branches (amplitude and phase mismatch between the different antenna or user front-ends) will be further presented.

- The aim of Chapter 5 is to describe our methodology based on a generic wireless communication system. The generic communication system will be similar to a single-antenna WLAN system based on OFDM. We will first present the structure of the frame based on which the different steps can be performed and afterwards describe all the steps in turn (burst detection and AGC, packet arrival time estimation, carrier frequency estimation, IQ imbalance estimation, carrier frequency offset and IQ imbalance compensation, tracking).

- In Chapter 6, the methodology proposed in the previous chapter will be applied to the emerging communication systems, in particular, taking into account the constraints from each system configuration. First, the multiple antenna WLAN IEEE 802.11n communication system will be considered. Second, the high-mobility long-term evolution of the 3GPP cellular system will be studied. The frame structure of both systems will be described, and the estimation/compensation algorithms necessary to mitigate the system specific effects will be presented.

2

New Air Interfaces

Because of the limited frequency bandwidth, on the one hand, and the potential limited power of terminal stations, on the other hand, spectral and power efficiencies of future communication systems should be as high as possible. New air interfaces need to be developed to meet the new user requirements.

The OFDM modulation has been selected for WLAN communication systems because it enables the low complexity equalization of the multi-path channel in the frequency domain (van Nee *et al.* 1999). In Section 2.1, we review the principles of OFDM and give a detailed mathematical description of the system. An interesting alternative air interface for WLANs, called signal-carrier frequency domain equalization (SC-FDE), is also introduced in Section 2.2. Multiple antenna technologies are finally combined to OFDM to improve the system reliability and capacity (see Section 2.3).

Cellular systems of the third generation (3G) are based on the recently emerged direct-sequence code division multiple access (DS-CDMA) technique (Ojanperä and Prasad 1998). DS-CDMA offers a potentially high system capacity and interesting networking abilities, such as soft hand-over. However, the DS-CDMA system suffers from interference (inter-symbol interference (ISI) and multi-user interference (MUI)) caused by multi-path propagation, leading to a high loss in spectral efficiency. We show in Section 2.4 that DS-CDMA can be combined with OFDM and SC-FDE to deal better with the multi-path propagation. As an alternative air interface for cellular systems, we also present the OFDMA and SC-FDMA air interfaces in Section 2.5.

2.1 Orthogonal frequency-division multiplexing

A block diagram of the OFDM-based communication system is illustrated in Figure 2.1. The main principle of OFDM is to consider the convolutive multi-path channel in the time-domain as a multiplicative frequency selective channel in the frequency-domain so that the complexity of the channel equalization can be significantly reduced. Therefore, an IFFT is performed at the transmitter on the transmitted symbols to move from the frequency domain to the time domain, where the resulting signal is convolved with the multi-path channel,

Digital Compensation for Analog Front-Ends François Horlin and André Bourdoux
© 2008 John Wiley & Sons, Ltd

and an FFT is performed at the receiver to move back to the frequency domain, where the frequency selective channel is equalized by means of complex coefficient multiplications. A detailed mathematical description of the OFDM system follows.

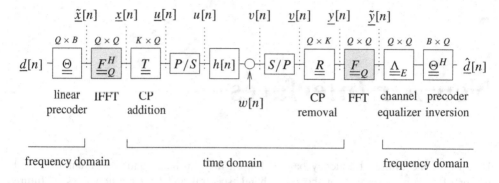

Figure 2.1 OFDM block diagram.

The information symbols, $d[n]$, which are assumed to be independent and of variance equal to σ_d^2, are first serial-to-parallel converted into blocks of B symbols, leading to the block sequence, $\underline{d}[n] := [d[nB], \ldots, d[(n+1)B-1]]^T$. The blocks $\underline{d}[n]$ are linearly pre-coded by means of a $Q \times B$ ($Q \geq B$) matrix, $\underline{\underline{\Theta}}$, which possibly introduces some redundancy:

$$\underline{\tilde{x}}[n] := \underline{\underline{\Theta}} \cdot \underline{d}[n] \tag{2.1}$$

The next operation involves the transformation of the frequency-domain block sequence, $\underline{\tilde{x}}[n]$, into the time-domain block sequence:

$$\underline{x}[n] = \underline{\underline{F}}_Q^H \cdot \underline{\tilde{x}}[n] \tag{2.2}$$

where $\underline{\underline{F}}_Q$ is the $Q \times Q$ FFT matrix, defined as:

$$\underline{\underline{F}}_Q := \frac{1}{\sqrt{Q}} [e^{-j2\pi(pq/Q)}]_{p,q=0,\ldots,Q-1} \tag{2.3}$$

Finally, the $K \times Q$ ($K \geq Q$) transmit matrix, $\underline{\underline{T}}$, possibly adds some transmit redundancy to the time-domain blocks:

$$\underline{u}[n] = \underline{\underline{T}} \cdot \underline{x}[n] \tag{2.4}$$

With $K = Q + L_{cp}$, $\underline{\underline{T}} := [\underline{\underline{I}}_{cp}^T, \underline{\underline{I}}_Q^T]^T$, where $\underline{\underline{I}}_{cp}$ consists of the last L rows of $\underline{\underline{I}}_Q$, $\underline{\underline{T}}$ adds redundancy in the form of a length-L_{cp} cyclic prefix (CP). The block sequence, $\underline{u}[n]$, is parallel-to-serial converted into the sequence of samples $u[n]$, such that $[u[nK], \ldots, u[(n+1)K-1]]^T := \underline{u}[n]$, and transmitted over the air at a rate $1/T$.

By setting $\underline{\underline{\Theta}} = \underline{\underline{I}}_Q$, we obtain the classical uncoded OFDM transmission scheme with Q parallel sub-carriers. OFDM has been adopted in several standards, including the digital audio broadcasting (DAB)/digital video broadcasting (DVB) standards in Europe, the high-speed digital subscriber line (DSL) standards for twisted-pair baseband communications, and the WLAN standards such as IEEE 802.11a and 802.11g. As an alternative to classical uncoded OFDM, linearly pre-coded OFDM is obtained by selecting $\underline{\underline{\Theta}}$ as a para-unitary matrix with $Q \geq B$, i.e., $\underline{\underline{\Theta}}^H \cdot \underline{\underline{\Theta}} = \underline{\underline{I}}_B$. The linear pre-coding can be either redundant ($Q > B$) (Wang and Giannakis 2003) or non-redundant ($Q = B$) (Liu et al. 2003). This section focuses on the classical uncoded OFDM transmission scheme. The next section will study the case of the non-redundant FFT linear pre-coder. In general, we assume in this text that redundancy is added in the form of a channel encoder applied in front of the linear pre-coder on the sequence of bits.

Adopting a discrete-time baseband equivalent model, the sampled received signal, $v[n]$, is a channel-distorted version of the transmitted user signal, which can be written as:

$$v[n] = \sum_{l=0}^{L} h[l]u[n-l] + w[n] \qquad (2.5)$$

where $h[l]$ is the sampled finite impulse response (FIR) channel of order L that models the frequency-selective multi-path propagation, including the effect of transmit/receive filters and the remaining synchronization error, and $w[n]$ is AWGN with variance σ_w^2.

The received sequence, $v[n]$, is serial-to-parallel converted into the corresponding block sequence, $\underline{v}[n] := [v[nK], \ldots, v[(n+1)K-1]]^T$. From the scalar input/output relationship in Equation (2.5), we can derive the corresponding block input/output relationship:

$$\underline{v}[n] = \underline{\underline{H}}[0] \cdot \underline{u}[n] + \underline{\underline{H}}[1] \cdot \underline{u}[n-1] + \underline{w}[n] \qquad (2.6)$$

where $\underline{w}[n] := [w[nK], \ldots, w[(n+1)K-1]]^T$ is the corresponding noise block sequence, $\underline{\underline{H}}[0]$ is a $K \times K$ lower triangular Toeplitz matrix with entries $[\underline{\underline{H}}[0]]_{p,q} = h[p-q]$, and $\underline{\underline{H}}[1]$ is a $K \times K$ upper triangular Toeplitz matrix with entries $[\underline{\underline{H}}[1]]_{p,q} = h[K+p-q]$ (see, e.g., Wang and Giannakis 2000, for a detailed derivation of the single-user case). The delay-dispersive nature of multi-path propagation gives rise to so-called inter-block interference (IBI) between successive blocks, which is modeled by the second term in Equation (2.6).

The $Q \times K$ receive matrix $\underline{\underline{R}}$ again removes the redundancy from the blocks, that is $\underline{y}[n] := \underline{\underline{R}} \cdot \underline{v}[n]$. With $\underline{\underline{R}} := [\underline{\underline{0}}_{Q \times L_{cp}}, \underline{\underline{I}}_Q]$, $\underline{\underline{R}}$ discards the length-L_{cp} cyclic prefix. The purpose of the transmit/receive pair is twofold. First, it allows for simple block-by-block processing by removing the IBI, that is $\underline{\underline{R}} \cdot \underline{\underline{H}}[1] \cdot \underline{\underline{T}} = \underline{\underline{0}}$, provided the cyclic prefix length to be at least equal to the channel order L. Second, it enables low-complexity frequency-domain processing by making the linear channel convolution to appear circulant to the received block. This results in a simplified block input/output relationship in the time-domain:

$$\underline{y}[n] = \underline{\underline{\dot{H}}} \cdot \underline{x}[n] + \underline{z}[n] \qquad (2.7)$$

where $\underline{\underline{\dot{H}}}$ is a circulant channel matrix, and $\underline{z}[n] := \underline{\underline{R}} \cdot \underline{w}[n]$ is the corresponding noise block sequence. Note that circulant matrices can be diagonalized by FFT operations:

$$\underline{\underline{F}}_Q \cdot \underline{\underline{\dot{H}}} \cdot \underline{\underline{F}}_Q^H = \underline{\underline{\Lambda}}_{\tilde{h}} \tag{2.8}$$

where $\underline{\underline{\Lambda}}_{\tilde{h}}$ is a diagonal matrix composed of the frequency-domain channel response.

In order to obtain an estimate, $\underline{\hat{d}}[n]$, of the transmitted symbol vector, $\underline{d}[n]$, based on the output vector, $\underline{y}[n]$, a usual approach is to use a linear equalizer optimized according to the minimum mean square error (MMSE) criterion. The MMSE linear detector minimizes the trace of the error auto-correlation matrix defined as:

$$\underline{\underline{R}}_{\epsilon\epsilon} := \mathcal{E}[\underline{\epsilon}[n] \cdot \underline{\epsilon}[n]^H] \tag{2.9}$$

in which $\underline{\epsilon}[n] := \underline{d}[n] - \underline{\hat{d}}[n]$ is the error vector. Note that the error auto-correlation matrix is independent of the block instant because the symbol and noise sequences are assumed to be stationary. At the output of the linear MMSE detector, the estimate of the vector of transmitted symbols is equal to (see Appendix A):

$$\underline{\hat{d}}[n] = \left(\frac{\sigma_w^2}{\sigma_d^2}\underline{\underline{I}}_Q + \underline{\underline{G}}^H \cdot \underline{\underline{G}}\right)^{-1} \cdot \underline{\underline{G}}^H \cdot \underline{y}[n] \tag{2.10}$$

in which $\underline{\underline{G}} := \underline{\underline{\dot{H}}} \cdot \underline{\underline{F}}_Q^H \cdot \underline{\underline{\Theta}}$ is the composite impulse response matrix (combination of Equations (2.1), (2.2) and (2.7)).

By remembering that the matrix $\underline{\underline{\dot{H}}}$ is circulant, and that the matrices $\underline{\underline{F}}_Q$ and $\underline{\underline{\Theta}}$ are (generally) unitary, the MMSE estimate (Equation (2.10)) becomes:

$$\underline{\hat{d}}[n] = \underline{\underline{\Theta}}^H \cdot \left(\frac{\sigma_w^2}{\sigma_d^2}\underline{\underline{I}}_Q + \underline{\underline{\Lambda}}_{\tilde{h}}^H \cdot \underline{\underline{\Lambda}}_{\tilde{h}}\right)^{-1} \cdot \underline{\underline{\Lambda}}_{\tilde{h}}^H \cdot \underline{\underline{F}}_Q \cdot \underline{y}[n] \tag{2.11}$$

Therefore, the MMSE linear detector can be seen as the succession of:

- an FFT to move to the frequency domain;
- the channel equalization in the frequency domain;
- the inverse of the linear precoder.

The right side of Figure 2.1 illustrates the block diagram of the receiver. The equalizer in the frequency domain is a diagonal matrix equal to:

$$\underline{\underline{\Lambda}}_{E,MMSE} = \left(\frac{\sigma_w^2}{\sigma_d^2}\underline{\underline{I}}_Q + \underline{\underline{\Lambda}}_{\tilde{h}}^H \cdot \underline{\underline{\Lambda}}_{\tilde{h}}\right)^{-1} \cdot \underline{\underline{\Lambda}}_{\tilde{h}}^H \tag{2.12}$$

It is well known that the linear MMSE detector reduces to the zero-forcing detector at very high values of the signal-to-noise ratio (SNR) (the term in σ_w^2/σ_d^2 is negligible) (Klein et al. 1996). In this case the channel is simply inverted in the frequency domain:

$$\underline{\underline{\Lambda}}_{E,ZF} = \underline{\underline{\Lambda}}_{\tilde{h}}^{-1} \tag{2.13}$$

NEW AIR INTERFACES

The principle of OFDM is to translate the convolution with the channel in the time domain to a multiplication in the frequency domain. Based on Equation (2.7), the output of the FFT at the receiver can be written as a function of the input of the IFFT at the transmitter as:

$$\tilde{y}[n] = \underline{\underline{\Lambda}}_{\tilde{h}} \cdot \tilde{x}[n] + \tilde{z}[n] \tag{2.14}$$

in which $\tilde{z}[n] := \underline{\underline{F}}_Q \cdot z[n]$ is the noise in the frequency domain. Therefore, OFDM creates Q independent parallel frequency domain channels, named the sub-carriers. Focusing on the sub-carrier q, Equation (2.14) reduces to ($q = 0, \ldots, Q - 1$):

$$\tilde{y}^q[n] = \tilde{h}^q \tilde{x}^q[n] + \tilde{z}^q[n] \tag{2.15}$$

where a^q is the element q of the vector $\underline{a}$ of size Q ($a = \tilde{x}, \tilde{y}, \tilde{z}$ or $\tilde{h}$, the channel in the frequency domain). Because the equalizer matrix given in Equation (2.12) is diagonal, the equalization can be seen as an operation per sub-carrier:

$$\lambda_E^q = \frac{\sigma_d^2 (\tilde{h}^q)^*}{\sigma_w^2 + \sigma_d^2 |\tilde{h}^q|^2} \tag{2.16}$$

The WLAN systems based on the IEEE 802.11a standard aim at high-speed communications in indoor environments (van Nee *et al.* 1999). They rely on OFDM to deal with the high multi-path components generated by the signal reflections on the walls of the rooms. The system parameters are summarized in Table 2.1. Only 48 sub-carriers are used to send data symbols. Pilots are placed on four sub-carriers to track the synchronization offsets and the phase noise. No signal is sent on the direct current (DC) sub-carrier nor on the remaining sub-carriers located on the side of the spectrum. Data rates up to 54 Mbps can be delivered to the mobile terminals.

Table 2.1 WLAN system parameters.

Carrier frequency	5 GHz (11a), 2.4 GHz (11b)
Signaling rate	20 MHz
Constellation	QPSK, 16QAM, 64QAM
Number of sub-carriers Q	64
Occupied sub-carriers	52
CP length L_{cp}	16
Channel code type	convolutional
Channel code rate	$\frac{1}{2}, \frac{2}{3}, \frac{3}{4}, 1$

2.2 Single-carrier with frequency domain equalization

Single-carrier with frequency domain equalization can be seen as a special case of linearly pre-coded OFDM (Czylwik 1997; Sari *et al.* 1995). An FFT matrix is used to pre-code the OFDM transmitted symbol blocks ($\underline{\underline{\Theta}} = \underline{\underline{F}}_Q$), that compensates for the IFFT at the transmitter of the classical OFDM system. The block diagram of the SC-FDE system is illustrated in Figure 2.2. The data symbols are convolved with the multi-path channel in the time domain. At the receiver, an FFT is performed to move to the frequency domain, where the multiplicative frequency selective channel can be compensated at a low complexity, and an IFFT is performed after the equalization to come back to the time domain.

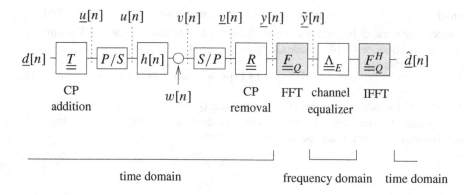

Figure 2.2 SC-FDE block diagram.

SC-FDE is an interesting alternative approach to OFDM. Both benefit from the same low-complexity multi-path channel equalization in the frequency domain. However, they feature different properties when they are implemented in actual wireless systems.

SC-FDE benefits from a lower peak-to-average power ratio (PAPR) than OFDM because no signal pre-coding with an IFFT is performed at the transmitter. Another difference is the computational complexity required at the transmitter and at the receiver. In the case of OFDM, one FFT/IFFT operator is performed at both sides of the link. In the case of SC-FDE, no FFT/IFFT operator is performed at the transmit side while two FFT/IFFT operators are performed at the receive side of the link. Interestingly, Falconer *et al.* (2002) encourages the use of SC-FDE in the uplink and OFDM in the downlink in order to reduce the constraints on the analog front-end and the processing complexity at the terminal.

The properties of OFDM and SC-FDE enable a dual extension of the two systems. In the case of OFDM, the time dispersive channel is seen in the frequency-domain as a set of parallel independent flat sub-channels. Using CSI at the transmitter, the constellation on each sub-channel can then be adapted to the quality, which is known as bit loading for adaptive OFDM. As a counter-part of adaptive OFDM, it has been proposed to use a decision feedback equalizer at the receive side of the SC-FDE system. The decision feedback equalizer makes use of the decisions already taken within the block to reduce the interference on the currently estimated symbols. It has been shown in (Benvenuto and Tomasin 2002; Louveaux *et al.* 2003) that the two approaches perform equally well.

We compare finally the performance of the OFDM and SC-FDE air interfaces in a WLAN communication system, operating in an indoor propagation environment (see Figure 2.3). The channel model proposed in (IEEE channel model 2004) has been used to generate the random channel realizations. The simulation parameters are compliant to the IEEE 802.11a standard parameters: a 5 GHz carrier frequency, a 20 MHz bandwidth, 64 sub-carriers,

52 occupied sub-carriers, a cyclic prefix length equal to 16. The modulation order and the channel coding rate are varying parameters.

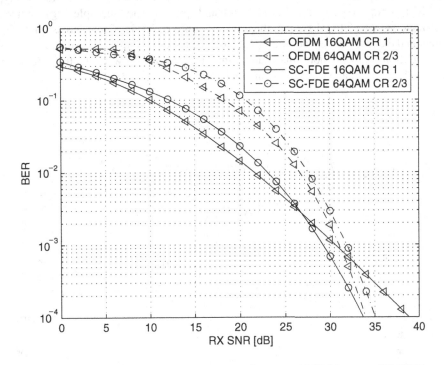

Figure 2.3 Performance of the single-antenna system (OFDM versus SC-FDE).

When no channel coding is used, the performance of SC-FDE crosses the performance of OFDM. OFDM outperforms SC-FDE at low SNR values but suffers rapidly from the channel fading while SC-FDE is able to benefit from the frequency diversity to compensate for it. For the same reason, it is better to increase the modulation order and to decrease proportionally the channel coding rate of the OFDM system at high SNR. As expected, this conclusion is also valid but less pronounced in the case of SC-FDE. It is interesting to note that SC-FDE offers a slightly higher spectral efficiency than OFDM since no guard bands are needed to limit the spectrum of the transmitted signal (the SC-FDE system relies on the transmitter pulse-shaping filter instead of on the modulation itself to limit the spectrum).

2.3 Multi-input multi-output OFDM

To meet the data rate and quality of service (QoS) requirements of emerging broadband communication systems, their spectral efficiency and link reliability should be considerably improved, which cannot be realized by using traditional single-antenna communication techniques. To achieve these goals, multiple-input multiple-output (MIMO) systems deploy multiple antennas at both ends of the wireless link to exploit the extra spatial dimension, besides the time and frequency dimensions (Foschini and Gans 1998; Gesbert *et al.* 2002; Raleigh and Cioffi 1998).

The OFDM system (given in Figure 2.1) is extended in Figure 2.4 to include two types of MIMO technique (space–time block coding and space-division multiplexing). Compared to the single-antenna OFDM system, the transmitter pre-codes the signal across N_T transmit antennas and the receiver decodes the signal based on the observation of N_R receive antennas. The sequence $h_{n_R n_T}[n]$ denotes the channel impulse response between the transmit antenna n_T ($n_T = 1, \ldots, N_T$) and the receive antenna n_R ($n_R = 1, \ldots, N_R$).

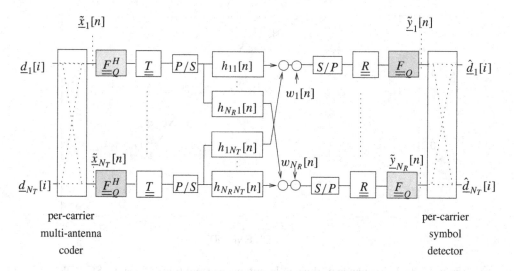

Figure 2.4 MIMO block diagram.

It has been shown in Section 2.1 that the OFDM system creates Q independent parallel fading channels in the frequency domain (named the sub-carriers). The MIMO techniques are applied on each sub-carrier q independently in order to keep the computational complexity as low as possible. The input–output relation (Equation (2.15)) is extended to MIMO systems as follows:

$$\underline{\tilde{y}}^q[n] = \underline{\underline{\tilde{H}}}^q \cdot \underline{\tilde{x}}^q[n] + \underline{\tilde{z}}^q[n] \tag{2.17}$$

in which the vector $\underline{\tilde{x}}^q[n]$ is a stacked version of the transmitted elements on the different transmit antennas:

$$\underline{\tilde{x}}^q[n] := \begin{bmatrix} \tilde{x}_1^q[n] & \cdots & \tilde{x}_{N_T}^q[n] \end{bmatrix}^T \tag{2.18}$$

and the vectors $\underline{\tilde{y}}^q[n]$ and $\underline{\tilde{z}}^q[n]$ are stacked versions of the received and noise elements on the different receive antennas:

$$\underline{\tilde{y}}^q[n] := \begin{bmatrix} \tilde{y}_1^q[n] & \cdots & \tilde{y}_{N_R}^q[n] \end{bmatrix}^T \tag{2.19}$$

$$\underline{\tilde{z}}^q[n] := \begin{bmatrix} \tilde{z}_1^q[n] & \cdots & \tilde{z}_{N_R}^q[n] \end{bmatrix}^T \tag{2.20}$$

and the matrix $\underline{\underline{\tilde{H}}}^q$ is the concatenation of the channel coefficients in the frequency domain:

$$\underline{\underline{\tilde{H}}}^q = \begin{bmatrix} \tilde{h}^q_{11} & \cdots & \tilde{h}^q_{1N_T} \\ \vdots & \ddots & \vdots \\ \tilde{h}^q_{N_R 1} & \cdots & \tilde{h}^q_{N_R N_T} \end{bmatrix} \quad (2.21)$$

The transmitted sequence $\underline{\tilde{x}}^q[n]$ is obtained by pre-coding the sequence of symbols $\underline{d}^q[i] := [d^q_1[i] \quad \cdots \quad d^q_{N_T}[i]]^T$. The multi-antenna pre-coder introduces potentially a change of rate (explaining why the index 'n' is not necessarily equal to the index 'i').

2.3.1 Space–time block coding

MIMO systems create $N_T N_R$ independently fading channels between the transmitter and the receiver, which allows an $N_T N_R$-fold diversity increase to be realized, where $N_T N_R$ is called the multi-antenna diversity gain.

Specifically, STC techniques exploit diversity and coding gains, by encoding the transmitted signals not only over the temporal domain but also over the spatial domain (Alamouti 1998; Tarokh et al. 1999, 1998). They bring diversity without requiring knowledge of the propagation channels at the transmitter. Within the class of space–time coding (STC) techniques, the space–time block coding (STBC) techniques, introduced by Alamouti (1998) for $N_T = 2$ transmit antennas, and later generalized by Tarokh et al. (1999) for any number of transmit antennas, are particularly appealing because they facilitate maximum likelihood (ML) detection with simple linear processing. However, these STBC techniques were originally designed for frequency-flat fading channels. Therefore, time-reversal STBC techniques, initially proposed by Lindskog and Paulraj (2000) for single-carrier serial transmission, have been combined with SC-FDE by Al-Dhahir (2001) and Zhou and Giannakis (2003) for signaling over frequency-selective fading channels.

In this section, we combine OFDM with STBC. For conciseness, we limit ourselves to the case of $N_T = 2$ transmit antennas (Alamouti 1998). However, the results can be easily extended to any number of antennas using the generalization proposed by Tarokh et al. (1999). The STBC coding is implemented by coding the two antenna streams across two time instants, as expressed in:

$$\underline{\tilde{x}}^q[n] = \underline{d}^q[i] \quad (2.22)$$

$$\underline{\tilde{x}}^q[n+1] = \underline{\underline{\chi}} \cdot (\underline{d}^q[i])^* \quad (2.23)$$

where $i = \lfloor n/2 \rfloor$, and:

$$\underline{\underline{\chi}} := \begin{bmatrix} 0 & -1 \\ 1 & 0 \end{bmatrix} \quad (2.24)$$

The transmitted block at time instant $n + 1$ from one antenna is the conjugate of the transmitted symbol at time instant n from the other antenna (with possible sign change). This property allows for deterministic transmit stream separation at the receiver, regardless of the underlying frequency selective channels.

At the receiver, the received signals corresponding to the two time instants are stacked on each other. Based on Equations (2.17), (2.22) and (2.23), we obtain:

$$\begin{bmatrix} \underline{\tilde{y}}^q[n] \\ (\underline{\tilde{y}}^q[n+1])^* \end{bmatrix} = \begin{bmatrix} \underline{\underline{\tilde{H}}}^q \\ (\underline{\underline{\tilde{H}}}^q)^* \cdot \underline{\underline{\chi}} \end{bmatrix} \cdot \underline{d}^q[i] + \begin{bmatrix} \underline{\tilde{z}}^q[n] \\ (\underline{\tilde{z}}^q[n+1])^* \end{bmatrix} \quad (2.25)$$

The first stage of the optimal ML receiver consists of a matched filter, implemented by multiplying the received vector with the Hermitian of the composite channel matrix:

$$\underline{\hat{d}}^q[i] = \begin{bmatrix} (\underline{\underline{\tilde{H}}}^q)^H & \underline{\underline{\chi}}^H (\underline{\underline{\tilde{H}}}^q)^T \end{bmatrix} \cdot \begin{bmatrix} \underline{\tilde{y}}^q[n] \\ (\underline{\tilde{y}}^q[n+1])^* \end{bmatrix} \quad (2.26)$$

Taking the definition of $\underline{\underline{\chi}}$ into account, it reduces to:

$$\underline{\hat{d}}^q[i] = \kappa \underline{d}^q[i] + \begin{bmatrix} (\underline{\underline{\tilde{H}}}^q)^H & \underline{\underline{\chi}}^H \cdot (\underline{\underline{\tilde{H}}}^q)^T \end{bmatrix} \cdot \begin{bmatrix} \underline{\tilde{z}}^q[n] \\ (\underline{\tilde{z}}^q[n+1])^* \end{bmatrix} \quad (2.27)$$

in which $\kappa := \sum_{n_T=1}^{2} \sum_{n_R=1}^{N_R} |\tilde{h}_{n_R n_T}^q|^2$. Therefore, the inter-antenna interference is cancelled out at the output of the matched filter so that the optimum ML receiver reduces to the matched filter stage. On the other hand, each transmitted symbol is multiplied with the sum of the square of the channel coefficients, bringing the desired diversity gain.

2.3.2 Space-division multiplexing

MIMO systems also create N_{min} parallel spatial pipes, enabling an N_{min}-fold capacity increase in rich scattering environments, where $N_{min} = \min\{N_T, N_R\}$ is called the spatial multiplexing gain (Foschini and Gans 1998; Gesbert *et al.* 2002; Raleigh and Cioffi 1998).

Specifically, space-division multiplexing (SDM) techniques exploit this spatial multiplexing gain, by simultaneously transmitting N_{min} independent information streams at the same frequency (Foschini 1996; Paulraj and Kailath 1994). The transmitted vector is a linearly pre-coded version of the symbol vector:

$$\underline{\tilde{x}}^q[n] = \underline{\underline{\Theta}}^q \cdot \underline{d}^q[i] \quad (2.28)$$

where $n = i$ and $\underline{\underline{\Theta}}^q$ is the SDM pre-coding matrix of size $N_T \times N_{min}$. Spatial multiplexing can be performed with or without channel knowledge at the transmitter (open-loop communication system or closed-loop communication system), so that the pre-coding matrix can potentially depend on the channel impulse response.

Open-loop communication system

When no CSI is available at the transmitter, the vector of symbols is often directly mapped onto the different transmit antennas, so that $\underline{\underline{\Theta}}^q = \underline{\underline{I}}_{N_T}$. In that case, the number of receive antennas should be at least equal to the number of transmit antennas ($N_{min} = N_T \leq N_R$). Because all antenna streams are transmitted simultaneously without any special pre-coding, they interfere with each other.

NEW AIR INTERFACES

Advanced receivers have to be foreseen to compensate for the inter-antenna interference and be able to benefit from the spatial multiplexing gain. The optimum detector, based on the ML criterion, suffers unfortunately from a prohibitive complexity because it implies an exhaustive search. An interesting alternative solution is to use a linear detector designed according to the MMSE criterion. At the output of the MMSE linear detector, the estimated vector of symbols is equal to (see Appendix A):

$$\hat{\underline{d}}^q[i] = \left(\frac{\sigma_w^2}{\sigma_d^2} \underline{\underline{I}}_{N_T} + (\underline{\underline{G}}^q)^H \cdot \underline{\underline{G}}^q \right)^{-1} \cdot (\underline{\underline{G}}^q)^H \cdot \underline{\tilde{y}}^q[n] \qquad (2.29)$$

in which $\underline{\underline{G}}^q := \underline{\underline{\tilde{H}}}^q \cdot \underline{\underline{\Theta}}^q$ is the composite channel impulse response (combination of Equations (2.17) and (2.28)).

Because the MMSE linear detector usually performs badly compared to the ML detector, it can be complemented with a successive interference cancellation receiver that iterates on the following steps:

- estimate one stream by subtracting the already estimated streams from the received signal;
- detect the estimated signal;
- reconstruct the contribution of the stream to the received signal.

The V-BLAST algorithm proposed in Foschini (1996) is essentially an ordered SIC, in which the stream with the highest SNR is decoded at every stage. The MMSE linear detector can be used to obtain an initial reliable estimate.

Closed-loop communication system

When the channel is known at the transmitter, significant performance gains can be obtained. The pre-coder (at the transmitter) and the decoder (at the receiver) can be jointly designed to optimize the information capacity (Scaglione *et al.* 1999), the error probability (Ding *et al.* 2003) and the symbol estimation MSE (Scaglione *et al.* 2002). In this section, we focus on the symbol estimation MSE optimization criterion. We assume initially that $N_{min} = N_T \leq N_R$, even if the final solution implies that some of the streams should be deactivated.

The receiver is first optimized by assuming a fixed pre-coding matrix $\underline{\underline{\Theta}}^q$. The optimum linear MMSE detector is given in Equation (2.29) and the MSE of each estimated symbol is found on the diagonal of the error auto-correlation matrix (see Appendix A):

$$\underline{\underline{R}}_{\epsilon\epsilon} = \left(\frac{1}{\sigma_d^2} \underline{\underline{I}}_{N_T} + \frac{1}{\sigma_w^2} (\underline{\underline{G}}^q)^H \cdot \underline{\underline{G}}^q \right)^{-1} \qquad (2.30)$$

Without any constraint, minimizing the MSE according to the pre-coding matrix leads to the trivial solution requiring an infinite transmit power. A reasonable constraint takes the limited power budget P into account. The problem becomes:

$$\min_{\underline{\underline{\Theta}}^q} \text{Tr}[\underline{\underline{R}}_{\epsilon\epsilon}] \quad \text{s.t.} \quad \sigma_d^2 \text{Tr}[\underline{\underline{\Theta}}^q \cdot (\underline{\underline{\Theta}}^q)^H] = P \qquad (2.31)$$

It is demonstrated by Scaglione et al. (2002) that the optimal pre-coding matrix is given by:

$$\underline{\underline{\Theta}}^q = \underline{\underline{V}}^q \cdot \underline{\underline{\Phi}}^q \tag{2.32}$$

in which $\underline{\underline{V}}^q$ is the matrix of eigenvectors of $(1/\sigma_w^2)(\underline{\underline{\tilde{H}}}^q)^H \cdot \underline{\underline{\tilde{H}}}^q$ and $\underline{\underline{\Phi}}^q$ is a diagonal matrix with complex coefficients ϕ_i^q on its diagonal. The coefficients ϕ_i^q are optimized according to the eigenvalues λ_i^q of the matrix $(1/\sigma_w^2)(\underline{\underline{\tilde{H}}}^q)^H \cdot \underline{\underline{\tilde{H}}}^q$. The problem reduces to:

$$\min_{\phi_i^q} \sum_{i=0}^{N_T-1} \frac{\sigma_d^2}{1+\sigma_d^2 \lambda_i^q |\phi_i^q|^2} \quad \text{s.t.} \quad \sigma_d^2 \sum_{i=0}^{N_T-1} |\phi_i^q|^2 = P \tag{2.33}$$

After the optimization under constraint, we obtain:

$$|\phi_i^q|^2 = \left(-\frac{1}{\sigma_d^2 \lambda_i^q} + \frac{1}{\sigma_d^2 \sqrt{\lambda_i^q}} \frac{P + \sum_{j=0}^{\bar{N}_T-1}(1/\lambda_j^q)}{\sum_{j=0}^{\bar{N}_T-1}(1/\sqrt{\lambda_j^q})} \right)^+ \tag{2.34}$$

where $(x)^+ := \max(x, 0)$ and $\bar{N}_T \leq N_T$ is such that $|\phi_j^q|^2 > 0$ for $j \in [0, \bar{N}_T - 1]$ and $|\phi_j^q|^2 = 0$ for all other j. Note that $\bar{N}_T$ is a function of the eigenvalues λ_q^i as well. It can be found iteratively by removing progressively the components with the highest indexes (associated with the smallest λ_q^i).

The IEEE 802.11n is the MIMO extension of the IEEE 802.11a/g standard for WLAN high throughput communications. It includes both the STBC and SDM techniques. Compared to the theoretical SDM closed-loop communication system presented in this text, the actual system works with the feedback of a partial knowledge of the channel impulse response (CIR) (Khaled et al. 2005; Love and Heath 2005).

Figure 2.5 illustrates the performance of the MIMO–OFDM air interfaces in a WLAN communication system, operating in an indoor propagation environment. The channel model (IEEE channel model 2004) has also been used to generate the random channel realizations. The simulation parameters are compliant with the IEEE 802.11n standard parameters: a 5 GHz carrier frequency, a 20 MHz bandwidth, 64 sub-carriers, 52 occupied sub-carriers, a cyclic prefix length equal to 16, a 16QAM modulation, no channel coding (to not hide the MIMO benefits).

Compared to the SISO system, the STBC 2×2 system (2 transmit and 2 receive antennas) benefits from the 3 dB receive array gain (less transmit power) and from the order 4 spatial diversity (improved reliability). The SDM systems (2×2 and 4×4) offer rather a capacity increase (factors 2 and 4, respectively) besides the receive array gain (3 and 6 dB, respectively). Note that it is not possible to observe the array gain in the figure because the BER is illustrated as a function of the received power (and not the transmit power). When a linear MMSE joint detector is used at the receiver to separate the transmit antenna streams, the bit error rate (BER) is worsening as the number of antennas increases. The BER of the SDM system applying a linear receiver is significantly worse than the BER of the single-input single-output (SISO) system. The successive interference canceller (SIC) in the open-loop SDM system or the channel state information (CSI)-based pre-coder in the closed-loop SDM system enables a similar significant BER improvement, showing that there exists a trade-off between the necessary computational complexity at the receiver and the information

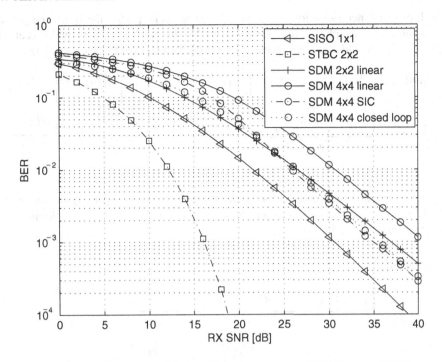

Figure 2.5 Performance of the multi-antenna system (STBC versus SDM).

feedback from the receiver to the transmitter. The performance of the SDM system applying a SIC receiver or a CSI-based pre-coder comes close to the performance of the single-antenna system.

2.3.3 Space-division multiple access

In systems where the access point or base station is fitted with multiple antennas, it is also possible to provide multiple access in the spatial domain. This multi-user technique is referred to as space-division multiple access (SDMA). SDMA can be used in both the forward and reverse links. In line-of-sight environments, it is easy to conceive SDMA as a form of beamsteering where different beams are aimed at different users, thereby separating their signals. It should be noted, however, that SDMA can also be applied in non-line-of-sight environments. In this case, wideband beamforming is required. Combined with OFDM, SDMA reduces to narrowband beamforming per sub-carrier just as in MIMO-OFDM systems (see Section 2.3). Finally, it is worth mentioning that, according to the number of antennas at the terminal, it is possible to design SDMA-multiple-input single-output (MISO) and SDMA-MIMO systems.

SDMA requires channel knowledge at the access point, for both the forward and reverse links. Many strategies are possible for SDMA (Vandenameele *et al.* 2000). We will assume that OFDM is being used so that the SDMA processing is performed per sub-carrier and that the terminals are SISO terminals.

With proper synchronization and a well-designed cyclic prefix, the following per sub-carrier system model applies in the forward link (K is the number of users, which is also the number of streams, and N_{AP} is the number of transmit antennas at the access point):

$$\underbrace{\begin{bmatrix} r_1^q[n] \\ \vdots \\ r_K^q[n] \end{bmatrix}}_{\underline{r}^q[n]} = \underbrace{\begin{bmatrix} h_{11}^q & \cdots & h_{1N_{AP}^q} \\ \vdots & \ddots & \vdots \\ h_{K1}^q & \cdots & h_{KN_{AP}^q} \end{bmatrix}}_{\underline{\underline{H}}^q} \cdot \underbrace{\begin{bmatrix} \theta_{11}^q & \cdots & \theta_{1K}^q \\ \vdots & \ddots & \vdots \\ \theta_{N_{AP}1}^q & \cdots & \theta_{N_{AP}K}^q \end{bmatrix}}_{\underline{\underline{\Theta}}^q} \cdot \underbrace{\begin{bmatrix} d_1^q[n] \\ \vdots \\ d_K^q[n] \end{bmatrix}}_{\underline{d}^q[n]} + \underbrace{\begin{bmatrix} w_1^q[n] \\ \vdots \\ w_K^q[n] \end{bmatrix}}_{\underline{w}^q[n]}$$

(2.35)

In this forward link model, a linear pre-coder $\underline{\underline{\Theta}}^q$ is used to map spatially the information symbols $\underline{d}^q[n]$ on the transmit antennas. Note that we must have $N_{AP} \geq K$. Similarly, we can write for the reverse link:

$$\underbrace{\begin{bmatrix} \hat{d}_1^q[n] \\ \vdots \\ \hat{d}_K^q[n] \end{bmatrix}}_{\hat{\underline{d}}^q[n]} = \underbrace{\begin{bmatrix} \xi_{11}^q & \cdots & \xi_{1N_{AP}}^q \\ \vdots & \ddots & \vdots \\ \xi_{K1}^q & \cdots & \xi_{KN_{AP}}^q \end{bmatrix}}_{\underline{\underline{\Xi}}^q} \cdot \underbrace{\begin{bmatrix} h_{11}^q & \cdots & h_{K1}^q \\ \vdots & \ddots & \vdots \\ h_{1N_{AP}}^q & \cdots & h_{KN_{AP}}^q \end{bmatrix}}_{\underline{\underline{H}}^{qT}} \cdot \underbrace{\begin{bmatrix} d_1^q \\ \vdots \\ d_K^q \end{bmatrix}}_{\underline{d}^q}[n]$$

$$+ \underbrace{\begin{bmatrix} \xi_{11}^q & \cdots & \xi_{1N_{AP}}^q \\ \vdots & \ddots & \vdots \\ \xi_{K1}^q & \cdots & \xi_{KN_{AP}}^q \end{bmatrix}}_{\underline{\underline{\Xi}}^q} [n] \cdot \underbrace{\begin{bmatrix} w_1^q[n] \\ \vdots \\ w_{N_{AP}}^q[n] \end{bmatrix}}_{\underline{w}^q[n]} \quad (2.36)$$

where $\underline{\underline{\Xi}}^q$ is a linear equalizer matrix applied at the access point to recover an estimate $\hat{\underline{d}}^q[n]$ of the transmitted symbols $\underline{d}^q[n]$. Under ideal assumptions, the reverse link channel is the transpose of the forward link channel.

Linear forward link strategies

Simple forward link strategies include the zero-forcing and MMSE precoders (Thoen *et al.* 2002):

$$\underline{\underline{\Theta}}_{ZF}^q = (\underline{\underline{H}}^q)^H \cdot (\underline{\underline{H}}^q \cdot (\underline{\underline{H}}^q)^H)^{-1} \quad (2.37)$$

$$\underline{\underline{\Theta}}_{MMSE}^q = (\underline{\underline{H}}^q)^H \cdot (\underline{\underline{H}}^q \cdot (\underline{\underline{H}}^q)^H + \lambda^q \underline{\underline{I}}_{N_{AP} \times N_{AP}})^{-1} \quad (2.38)$$

In Equation (2.38), λ^q is a Lagrange parameter that needs to be computed to meet a transmit power constraint of the form

$$\text{Tr}[\underline{\underline{\Theta}}^q \cdot \underline{\underline{R}}_{dd}^q (\underline{\underline{\Theta}}^q)^H] = P_T^q \quad (2.39)$$

where P_T^q is the allowed transmit power per sub-carrier and

$$\underline{\underline{R}}_{dd}^q = \mathcal{E}[\underline{d}^q \cdot (\underline{d}^q)^H] \quad (2.40)$$

NEW AIR INTERFACES

is the auto-correlation matrix of the transmitted symbol vector. The zero-forcing pre-coder perfectly cancels the inter-stream interference (multi-user interference) but it is known to suffer from transmit power increase, which is analogous to noise enhancement in zero-forcing receivers. This motivates the use of the MMSE pre-coder which tolerates a small amount of MUI and significantly reduces the power increase problem (Thoen *et al.* 2002). Nonlinear strategies are also possible in the forward link, such as the Tomlinson–Harashima pre-coding (Harashima and Mikayawa 1972).

Reverse link strategies

In the reverse link, simple linear receivers can also apply the zero-forcing or MMSE criterion. The reverse link zero-forcing and MMSE receivers are the transpose of those derived for the forward link. More complex, but non-linear receivers can be designed such as the ordered successive interference receiver (Vandenameele *et al.* 2000), the sphere decoder (Hochwald and ten Brink 2003) or the maximum likelihood receiver.

2.4 Code-division multiple access

2.4.1 Direct-sequence code-division multiple access

Cellular systems of the third generation (3G) are based on the DS-CDMA technology (Ojanperä and Prasad 1998) to separate the users based on a set of codes. The DS-CDMA technology is characterized by two principles:

- The user signals are spread with specific orthogonal binary codes to separate the users within a single cell. The DS-CDMA signals occupy a larger bandwidth (assuming that the symbol rate is fixed) so that the propagation channels become frequency selective (hence, the name wideband-CDMA). Typical Rake receivers exploit the resulting multi-path diversity to improve the communication reliability. In the downlink, an orthogonal set of codes is often used because the user signals are received synchronously at each mobile terminal. In the uplink, a quasi-orthogonal set of codes is often used because the user signals are received asynchronously at the base station (when no specific user pre-synchronization mechanism is foreseen). DS-CDMA enables asynchronous communications in the uplink.

- In a cellular system, the base stations can be further distinguished by a scrambling code and can use the same carrier frequency (universal frequency reuse). A mobile terminal located at the boundary between two cells can separate the signals coming from the two different base stations by correlating the composite received signal with each scrambling code (the interfering signal is averaged out). This property enables a soft hand-over between two neighboring cells because the terminal can initiate a communication with the new base station while still communicating with the former one, even if the analog front-end is in one given configuration.

Even if DS-CDMA offers interesting advantages (multi-path diversity, asynchronous uplink communications, soft hand-over), a DS-CDMA system suffers from interference caused by the multi-path propagation and by the user asynchronism (inter-symbol, multi-user and inter-cell interference), leading to a significant loss of performance in typical outdoor environments. For this reason, the spectral efficiency of 3G DS-CDMA communication systems is very low.

In order to enable the design of low-complexity transceivers that can cope with multi-path channels while still benefiting from the good properties of DS-CDMA, next-generation cellular systems could combine the DS-CDMA accessing scheme with the OFDM modulation (Section 2.1) or with the SC-FDE modulation (Section 2.2). As indicated in Section 2.2, it is advantageous to select OFDM in the downlink and SC-FDE in the uplink in order to reduce the computational complexity and the constraints on the power amplifier at the mobile terminal. This results in the multi-carrier code-division multiple access (MC-CDMA) (combination of DS-CDMA and OFDM) and cyclic prefix code-division multiple access (CP-CDMA) (combination of DS-CDMA and SC-FDE) air interfaces.

2.4.2 Multi-carrier code-division multiple access

MC-CDMA first performs classical DS-CDMA symbol spreading, followed by OFDM modulation, such that the information symbols are spread across the different sub-carriers located at different frequencies and characterized by different fading coefficients (Fazel 1993; Fazel et al. 1995; Kaiser 2002; Yee et al. 1993). Therefore, the MC-CDMA system is able to benefit from the channel frequency diversity.

The downlink of a communication system based on MC-CDMA is illustrated in Figure 2.6. The base station transmits simultaneously K signals to different mobile terminals. At each mobile terminal, one symbol stream is estimated based on the composite received signal (symbol stream k in Figure 2.6).

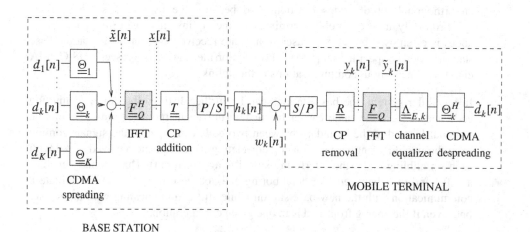

Figure 2.6 MC-CDMA block diagram (downlink, terminal k).

NEW AIR INTERFACES

At the transmitter base station, the symbol vectors $\underline{d}_k[n]$ of size B are spread with the spreading matrices $\underline{\underline{\Theta}}_k$ and added together to produce the vector of chips $\underline{\tilde{x}}[n]$:

$$\underline{\tilde{x}}[n] = \sum_{k=1}^{K} \underline{\underline{\Theta}}_k \cdot \underline{d}_k[n] \qquad (2.41)$$

With $Q = NB$ and N the spreading code length, the $Q \times B$ spreading matrix $\underline{\underline{\Theta}}_k$ is defined as:

$$\underline{\underline{\Theta}}_k := \underline{c}_k \otimes \underline{\underline{I}}_B \qquad (2.42)$$

where $\underline{c}_k := [c_k[0] \cdots c_k[N-1]]^T$ is the $N \times 1$ code vector of user k ($k = 1, \ldots, K$). The codes are orthogonal when $\underline{c}_k^H \cdot \underline{c}_l = \delta_{kl}$, or equivalently when $\underline{\underline{\Theta}}_k^H \cdot \underline{\underline{\Theta}}_l = \delta_{kl} \underline{\underline{I}}_B$.

The vector $\underline{\tilde{x}}[n]$ is generated in the frequency domain at the base station. An IFFT is performed to move to the time domain, where redundancy is added in the form of a cyclic prefix. The resulting signal is transmitted over the air and received at the mobile terminal k through a single propagation channel. After cyclic prefix removal and the FFT to move back to the frequency domain, the vector of symbols $\underline{d}_k[n]$ is estimated based on the vector $\underline{\tilde{y}}_k[n]$. A single-user detector is usually used, that consists of a chip equalizer $\underline{\underline{\Lambda}}_{E,k}$ to estimate the vector of chips $\underline{\tilde{x}}[n]$, and of a code correlator $\underline{\underline{\Theta}}_k^H$ to separate the user signals. The chip equalizer is often designed according to the zero-forcing criterion ($\underline{\underline{\Lambda}}_{E,k} = \underline{\underline{\Lambda}}_{\tilde{h}_k}^{-1}$) because the MMSE equalizer which implies the inversion of a size-Q square matrix cannot be implemented at a low complexity when the input symbols are correlated (which is the case at the output of the DS-CDMA modulator). When the DS-CDMA codes are orthogonal, the estimated symbol vector reduces to:

$$\underline{\hat{d}}_k[n] = \underline{d}_k[n] + \underline{\underline{\Theta}}_k^H \cdot \underline{\underline{\Lambda}}_{\tilde{h}_k}^{-1} \cdot \underline{\tilde{z}}[n] \qquad (2.43)$$

based on Equations (2.13), (2.14) and (2.41).

2.4.3 Cyclic-prefix code-division multiple access

CP-CDMA first performs classical DS-CDMA symbol spreading, followed by SC-FDE modulation, such that the information symbols are spread across the different single-carrier block transmission (SCBT) sub-channels (Baum *et al.* 2002; Vollmer *et al.* 2001).

The uplink of a communication system based on CP-CDMA is illustrated in Figure 2.7. The mobile terminals transmit synchronously a cyclic signal to the base station (as opposed to the classical asynchronous DS-CDMA mode). Because the user signals are received at the base station through different propagation channels, the orthogonality between the user signals cannot be restored based on a simple channel equalization. Advanced joint detection techniques have to be considered that detect jointly the user symbol blocks transmitted over different propagation channels. The complexity of the joint detector can still be significantly reduced by exploiting the circulant properties of the channel matrices (Vollmer *et al.* 2001).

At each mobile terminal, the symbol vector $\underline{d}_k[n]$ is spread with the spreading matrix $\underline{\underline{\Theta}}_k$ to produce the vector of chips $\underline{x}_k[n]$:

$$\underline{x}_k[n] = \underline{\underline{\Theta}}_k \cdot \underline{d}_k[n] \qquad (2.44)$$

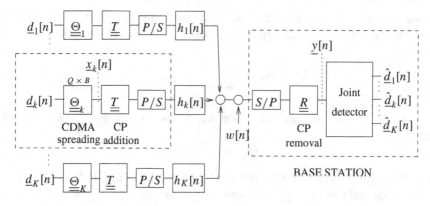

Figure 2.7 CP-CDMA block diagram.

With $Q := BN$ and N the spreading code length, the $Q \times B$ spreading matrix, $\underline{\underline{\Theta}}_k$ is generally defined as:

$$\underline{\underline{\Theta}}_k := \underline{\underline{I}}_B \otimes \underline{c}_k \qquad (2.45)$$

where $\underline{c}_k$ is the $N \times 1$ code vector of user k ($k = 1, \ldots, K$). As opposed to the MC-CDMA system in which the chips associated to one symbol are distributed over the set of sub-carriers in the frequency domain to benefit from the channel frequency diversity, the chips associated to one symbol in a CP-CDMA system are set on adjacent sub-channels in the time domain.

Redundancy is added in the form of a cyclic prefix to the vector $\underline{x}_k[n]$ and the result is sent to the base station. The base station receives the sum of the channel distorted versions of the transmitted signals plus additive noise. Removing the cyclic prefix from the received vector results in the time-domain vector $\underline{y}[n]$:

$$\underline{y}[n] = \sum_{k=1}^{K} \underline{\underline{\dot{H}}}_k \cdot \underline{x}_k[n] + \underline{z}[n] \qquad (2.46)$$

which is the multi-user extension of Equation (2.7).

In order to design the multi-user joint detector for the CP-CDMA system, a generalized matrix model is built that links the vector of input symbols to the vector of received samples. The generalized input/output matrix model that relates the symbol vector defined as:

$$\underline{d}[n] := \begin{bmatrix} \underline{d}_1[n]^T & \cdots & \underline{d}_K[n]^T \end{bmatrix}^T \qquad (2.47)$$

to the received vector $\underline{y}[n]$ and noise vector $\underline{z}[n]$, is given by:

$$\underline{y}[n] = \underline{\underline{H}} \cdot \underline{\underline{\Theta}} \cdot \underline{d}[n] + \underline{z}[n] \qquad (2.48)$$

where the channel matrix is equal to:

$$\underline{\underline{H}} := \begin{bmatrix} \underline{\underline{\dot{H}}}_1 & \cdots & \underline{\underline{\dot{H}}}_K \end{bmatrix} \qquad (2.49)$$

NEW AIR INTERFACES

and the spreading matrix is equal to:

$$\underline{\underline{\Theta}} := \begin{bmatrix} \underline{\underline{\Theta}}_1 & \cdots & \underline{\underline{0}}_{Q \times B} \\ \vdots & \ddots & \vdots \\ \underline{\underline{0}}_{Q \times B} & \cdots & \underline{\underline{\Theta}}_K \end{bmatrix} \qquad (2.50)$$

A factorization of the matrix $\underline{\underline{H}} \cdot \underline{\underline{\Theta}}$ serves as the basis of the joint detector simplification. Based on Equations (2.49) and (2.50), we have that:

$$\underline{\underline{H}} \cdot \underline{\underline{\Theta}} = \begin{bmatrix} \underline{\dot{\underline{H}}}_1 \cdot \underline{\underline{\Theta}}_1 & \cdots & \underline{\dot{\underline{H}}}_K \cdot \underline{\underline{\Theta}}_K \end{bmatrix} \qquad (2.51)$$

which can be interestingly reorganized into:

$$\underline{\underline{H}} \cdot \underline{\underline{\Theta}} = \underline{\underline{\Psi}} \cdot \underline{\underline{\Upsilon}} \qquad (2.52)$$

where $\underline{\underline{\Upsilon}}$ is a permutation matrix of size KB whose role is to reorganize the columns of the initial matrix $\underline{\underline{H}} \cdot \underline{\underline{\Theta}}$ according to the symbol and user indexes successively, and:

$$\underline{\underline{\Psi}} := \begin{bmatrix} \underline{\underline{\psi}}[1] & \cdots & \underline{\underline{\psi}}[B] \end{bmatrix} \qquad (2.53)$$

in which the matrix $\underline{\underline{\psi}}[1]$ of size $Q \times K$ is composed of the K user channel responses convolved with the user codes, and each matrix $\underline{\underline{\psi}}[b]$ of size $Q \times K$ is a $(b-1)N$ cyclic rotation of the matrix $\underline{\underline{\psi}}[1]$ ($b = 2, \ldots, B$).

The block circulant matrix $\underline{\underline{\Psi}}$ can be decomposed according to Vollmer et al. (2001) into:

$$\underline{\underline{\Psi}} = \underline{\underline{F}}_{(N)}^H \cdot \underline{\underline{\tilde{\Psi}}} \cdot \underline{\underline{F}}_{(K)} \qquad (2.54)$$

where the matrices $\underline{\underline{F}}_{(N)}$ and $\underline{\underline{F}}_{(K)}$ are block Fourier transforms, defined as:

$$\underline{\underline{F}}_{(n)} := \underline{\underline{F}}_B \otimes \underline{\underline{I}}_n \qquad (2.55)$$

where $\underline{\underline{F}}_B$ is the orthogonal Fourier transform matrix of size B and $\underline{\underline{I}}_n$ is the identity matrix of size n (n equal to N or K). The inner matrix $\underline{\underline{\tilde{\Psi}}}$ is equal to:

$$\underline{\underline{\tilde{\Psi}}} := \begin{bmatrix} \underline{\underline{\tilde{\psi}}}[1] & \cdots & \underline{\underline{0}}_{N \times K} \\ \vdots & \ddots & \vdots \\ \underline{\underline{0}}_{N \times K} & \cdots & \underline{\underline{\tilde{\psi}}}[B] \end{bmatrix} \qquad (2.56)$$

where the block diagonal is found by dividing the result of the product $\underline{\underline{F}}_{(N)} \cdot \underline{\underline{\psi}}[1]$ in blocks $\underline{\underline{\tilde{\psi}}}[b]$ of size $N \times K$.

We obtain finally:

$$\underline{\underline{H}} \cdot \underline{\underline{\Theta}} = \underline{\underline{F}}_{(N)}^H \cdot \underline{\underline{\tilde{\Psi}}} \cdot \underline{\underline{F}}_{(K)} \cdot \underline{\underline{\Upsilon}} \qquad (2.57)$$

and the matrix model (2.48) becomes:

$$\underline{y}[n] = \underline{\underline{F}}_{(N)}^H \cdot \underline{\underline{\tilde{\Psi}}} \cdot \underline{\underline{F}}_{(K)} \cdot \underline{\underline{\Upsilon}} \cdot \underline{d}[n] + \underline{z}[n] \qquad (2.58)$$

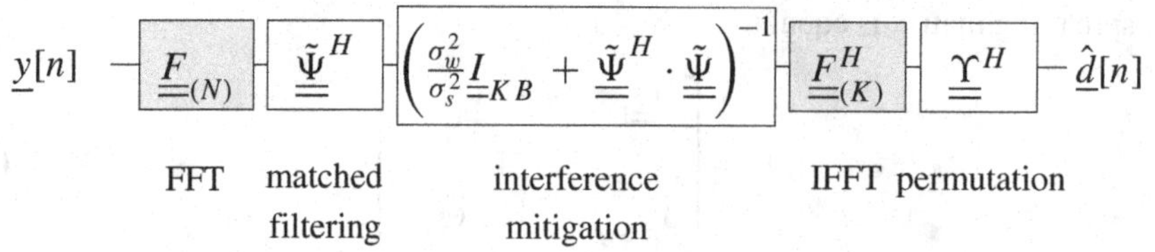

$$\underline{y}[n] \longrightarrow \boxed{\underline{\underline{F}}_{(N)}} \longrightarrow \boxed{\underline{\underline{\tilde{\Psi}}}^H} \longrightarrow \boxed{\left(\frac{\sigma_w^2}{\sigma_s^2}\underline{\underline{I}}_{=KB} + \underline{\underline{\tilde{\Psi}}}^H \cdot \underline{\underline{\tilde{\Psi}}}\right)^{-1}} \longrightarrow \boxed{\underline{\underline{F}}_{(K)}^H} \longrightarrow \boxed{\underline{\underline{\Upsilon}}^H} \longrightarrow \underline{\hat{d}}[n]$$

FFT   matched   interference   IFFT permutation
           filtering   mitigation

Figure 2.8 CP-CDMA joint detector.

The optimal solution to detect the CP-CDMA transmitted signals is to estimate jointly the symbol blocks of the different users within the transmitted vector, $\underline{d}[n]$, based on the received block, $\underline{y}[n]$ (multi-user joint detection). The optimum linear joint detector according to the MMSE criterion is computed in Appendix A. At the output of the MMSE multi-user detector, the estimate of the transmitted vector is:

$$\underline{\hat{d}}[n] = \underline{\underline{\Upsilon}}^H \cdot \underline{\underline{F}}_{(K)}^H \cdot \left(\frac{\sigma_w^2}{\sigma_s^2}\underline{\underline{I}}_{=KB} + \underline{\underline{\tilde{\Psi}}}^H \cdot \underline{\underline{\tilde{\Psi}}}\right)^{-1} \cdot \underline{\underline{\tilde{\Psi}}}^H \cdot \underline{\underline{F}}_{(N)} \cdot \underline{y}[n] \qquad (2.59)$$

which has been obtained by using the fact that the matrices $\underline{\underline{F}}_{(N)}$, $\underline{\underline{F}}_{(K)}$ and $\underline{\underline{\Upsilon}}$ are unitary. The MMSE linear joint detector decomposes successively into the following operations (see Figure 2.8):

(i) the FFT to move to the frequency domain;

(ii) the matched filtering $\underline{\underline{\tilde{\Psi}}}^H$ in the frequency domain;

(iii) the mitigation of the multi-user interference by multiplication with the inner matrix;

(iv) the IFFT to go back to the time domain;

(v) the permutation $\underline{\underline{\Upsilon}}$ to arrange the result according to the user and symbol indexes successively.

The matrix $\underline{\underline{\tilde{\Psi}}}$ defined in Equation (2.57) is block diagonal so that the computation of the MMSE joint detector involves only the inversion of B square auto-correlation matrices of size K.

Similarly to what has been observed for the SDM multi-antenna system, the MMSE linear joint detector performs generally poorly compared to the optimal ML detector. It can be complemented with a SIC, that progressively removes the interference caused by the already estimated symbols from the received signal.

2.5 Frequency-division multiple access

2.5.1 Orthogonal frequency-division multiple access

Next-generation outdoor communication systems will rely on the orthogonal frequency-division multiple access (OFDMA) technique to separate the users in the frequency domain

NEW AIR INTERFACES

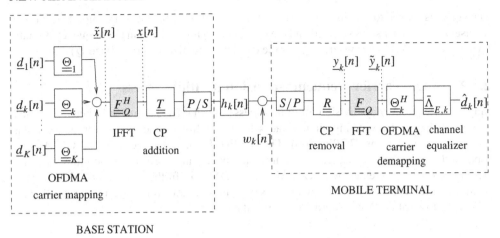

Figure 2.9 OFDMA block diagram.

(3GPP 2006; IEEE 2005). The OFDMA air interface relies on OFDM to create a set of orthogonal sub-channels that can be shared among the different users.

The downlink of the OFDMA-based communication system is illustrated in Figure 2.9. At the transmitter base station, the size-B symbol vectors $\underline{d}_k[n]$ are mapped with the OFDMA sub-carrier mapping matrices $\underline{\underline{\Theta}}_k$ onto the user specific sets of sub-carriers and added together to produce the composite vector of symbols $\underline{\tilde{x}}[n]$ in the frequency domain. We distinguish two kinds of sub-carrier repartition.

1. In the case of a localized transmission, the sub-carriers allocated to each user are adjacent on the spectrum. The OFDMA sub-carrier mapping matrices are typically defined as:

$$\underline{\underline{\Theta}}_k := \underline{c}_k \otimes \underline{\underline{I}}_B \qquad (2.60)$$

   where $\underline{c}_k$ is a column vector defining the position of the sub-carriers allocated to user k (equal to 1 at the position allocated to the user k and 0 elsewhere). This mode is often used at low-mobility conditions because it is possible to share the sub-carriers among the users optimally based on channel information obtained through feedback at the transmitter. Because each part of the spectrum can be allocated to the user having the highest channel gain at this place, it benefits from multi-user diversity.

2. In the case of a distributed transmission, the sub-carriers allocated to each user are spread over the whole spectrum. The OFDMA sub-carrier mapping matrices are typically defined as:

$$\underline{\underline{\Theta}}_k := \underline{\underline{I}}_B \otimes \underline{c}_k \qquad (2.61)$$

   where $\underline{c}_k$ is the same column vector defining the position of the sub-carriers allocated to user k. This mode is often used at high-mobility conditions because it is difficult to rely on any relevant channel information to share the sub-carriers among the users in that case. Because each user signal is spread over the whole spectrum, the system can benefit from frequency diversity.

The signal is moved to the time domain and transmitted over the air. At the mobile terminal, the user k sub-carrier set is selected and the user channel is equalized in the frequency domain ($\underline{\bar{\underline{\Lambda}}}_{E,k} := \underline{\underline{\Theta}}_k^H \cdot \underline{\underline{\Lambda}}_{E,k} \cdot \underline{\underline{\Theta}}_k$ where $\underline{\underline{\Lambda}}_{E,k}$ is the diagonal equalizer matrix of size Q).

2.5.2 Single-carrier frequency-division multiple access

Single-carrier frequency-division multiple access (SC-FDMA) has recently been identified as one of the most promising air interfaces for the uplink of the next-generation cellular communication systems (3GPP 2006). The SC-FDMA air interface relies on OFDMA to separate the users in the frequency domain, and hence can benefit from the multi-user and frequency diversity created with OFDMA. SC-FDMA is further characterized by a pre-coding of the vector of symbols with a discrete Fourier transform (DFT) of size $B < Q$ applied in front of OFDMA in order to reduce the PAPR.

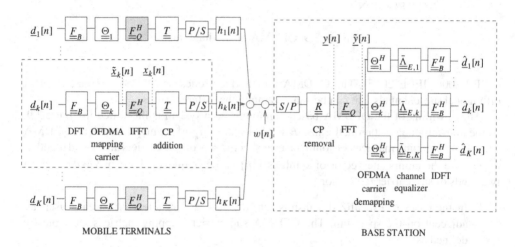

Figure 2.10 SC-FDMA block diagram.

Figure 2.10 gives the block diagram of the uplink of the SC-FDMA system. At each mobile terminal, the vector of symbols is pre-coded with a DFT and the result is mapped on the user allocated sub-carrier set. Both localized and distributed sub-carrier mappings can be realized. The signal is moved to the time domain and transmitted through the user-specific channel to the base station. The receiver base station selects the sub-carrier set corresponding to each user and equalizes the corresponding user channel in the frequency domain. Finally, the symbol vector estimate is obtained by inverting the DFT pre-coding.

The Third Generation Partnership Project (3GPP) standardization committee has recently started a new study group to define the long-term evolution of the cellular communication systems (3GPP long-term evolution (LTE) study group). The new system should be competitive over the next 10 years. Therefore, data rates up to 100 Mbps in the downlink and up to 20 Mbps in the uplink should be offered to the mobile terminals at high vehicular speeds. In the downlink, OFDMA has been selected because it offers a high flexibility in the allocation of the resources to the users. The uplink will rather rely on SC-FDMA to

Table 2.2 3GPP LTE system parameters.

Carrier frequency			2 GHz		
Bandwidth	1.25	2.5	5	10	20 MHz
Signaling rate	1.92	3.84	7.68	15.36	30.72 MHz
Number of sub-carriers	128	256	512	1024	2048
Occupied sub-carriers	76	151	301	601	1201
Short CP	10	20	40	80	160
Long CP	32	64	128	256	512
DL constellation		QPSK,	16QAM,	64QAM	
UL constellation		QPSK,	8PSK,	16QAM	
Channel code type			To be defined		
Channel code rate			To be defined		

reduce the constraints on the terminal analog front-end. A feasibility study of the technology has been performed in the 3GPP LTE study group and has shown that the targets could be met. Table 2.2 summarizes the system parameters. Several bandwidths are supported (from 1.25 to 20 MHz). The FFT size varies proportionally to the bandwidth so that the robustness against the mobility is maintained for all bandwidths (the symbol block duration and the inter-sub-carrier spacing is constant). Fewer than three-fifth of the sub-carriers are used effectively (the DC sub-carrier and the other sub-carriers on the sides of the spectrum are set to zero). The cyclic prefix length can be varied according to the actual channel duration (short or long cyclic prefix). Physical blocks of adjacent sub-carriers are defined. In the case of localized transmissions, each physical block is allocated to one user. In the case of distributed transmissions, the physical blocks are divided among the users (each user is allocated one sub-carrier from each physical block).

We evaluate finally the performance of the code division multiple access (CDMA) and frequency-division multiple access (FDMA) air interfaces in a cellular communication system, which operates in an outdoor sub-urban macro-cell propagation environment. The 3GPP TR25.996 geometrical spatial channel model (3GPP channel model 2003) has been considered to generate random channel realizations. Rather than comparing all possible modes for the two link directions, we consider only the relevant modes for each direction. For the uplink, the single carrier (SC) modes (CP-CDMA and SC-FDMA) are selected because they exhibit a smaller PAPR than the multi-carrier (MC) modes, which leads to increased terminal power efficiency. For the downlink, the MC modes (MC-CDMA and OFDMA) are the preferred modulation schemes, since they incur only a single FFT operation at the receiver side which also leads to reduced terminal complexity.

The modulation parameters are selected according to the 3GPP LTE standard: a 2 GHz carrier frequency, a 5 MHz bandwidth, 512 sub-carriers, 301 occupied sub-carriers, a cyclic prefix length equal to 40, a 16QAM constellation. We keep the widely used convolutional channel encoder and the bit interleaver defined in the IEEE 802.11a/n standard. In the case of CDMA (MC-CDMA/CP-CDMA), the user signals are spread by periodic Walsh–Hadamard codes of length 16, which are overlayed with an aperiodic Gold code for scrambling. In the case of FDMA (OFDMA/SC-FDMA), physical blocks of 16 sub-carriers are formed that can be allocated in a localized or distributed manner to the users.

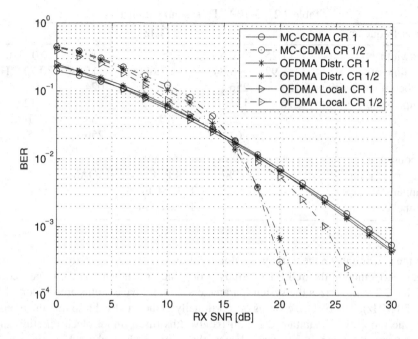

Figure 2.11 Performance in the downlink (MC-CDMA versus OFDMA).

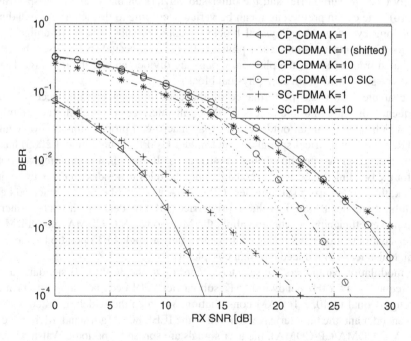

Figure 2.12 Performance in the uplink (uncoded CP-CDMA versus uncoded SC-FDMA with a distributed allocation).

Figure 2.11 compares the performance of MC-CDMA and OFDMA with a localized or distributed sub-carrier allocation in the downlink. Because both systems are orthogonal, their performance is independent of the number of users. When no channel coding is considered in the system, MC-CDMA and OFDMA perform equally well on average (both suffer from channel fading). Conversely, coded MC-CDMA and coded OFDMA with a distributed sub-carrier allocation significantly outperform OFDMA with a localized sub-carrier allocation because they are better able to benefit from the frequency diversity in the channels to compensate for the channel fading.

Figure 2.12 compares the performance of uncoded CP-CDMA and SC-FDMA with a distributed sub-carrier allocation in the uplink. Compared to SC-FDMA, CP-CDMA enables the system to benefit from the frequency diversity in the channels to compensate for the channel fading but suffers from the interference caused by the multi-path propagation (ISI and MUI). When 10 users are active in the system, SC-FDMA outperforms CP-CDMA at low signal-to-noise ratio (SNR) values (the interference is the dominant effect) while CP-CDMA outperforms SC-FDMA at high SNR values (the channel fading is the dominant effect). The performance of the CP-CDMA system is significantly better when an SIC equalizer complements the linear MMSE equalizer. However, the improvement does not enable the system to reach the performance of the single-user system due to the error propagation mechanism (see the single user curve shifted by 10 dB). It is interesting to note that no guard bands have been assumed in the CP-CDMA system so that its capacity is higher than in the SC-FDMA system.

References

3GPP (2006) 3GPP TR 25.814, 3rd generation partnership project; technical specification group radio access network; physical layer aspects for evolved UTRA.

3GPP channel model (2003) TR25.996 v6.1.0, spatial channel model for multiple input multiple output (MIMO) simulations.

Al-Dhahir, N. (2001) Single-carrier frequency-domain equalization for space–time block-coded transmissions over frequency-selective fading channels. *IEEE Communications Letters* **5**(7), 304–306.

Alamouti, S. (1998) A simple transmit diversity technique for wireless communications. *IEEE Journal on Selected Areas in Communications* **16**(8), 1451–1458.

Baum, K., Thomas, T., Vook, F. and Nangia, V. (2002) Cyclic-prefix CDMA: An improved transmission method for broadband DS-CDMA cellular systems. *IEEE Proceedings of Wireless Communications and Networking Conference*, vol. 1, pp. 183–188.

Benvenuto, N. and Tomasin, S. (2002) On the comparison between OFDM and single carrier modulation with a DFE using a frequency-domain feedforward filter. *IEEE Transactions on Communications* **50**(6), 947–955.

Czylwik, A. (1997) Comparison between adaptive OFDM and single carrier modulation with frequency domain equalization. *IEEE Proceedings of Vehicular Technology Conference*, pp. 865–869.

Ding, Y., Davidson, N., Luo, Z. and Wong, K. (2003) Minimum BER block precoders for zero-forcing equalization. *Signal Processing* **51**, 2410–2423.

Falconer, D., Ariyavisitakul, S., Benyamin-Seeyar, A. and Eidson, B. (2002) Frequency domain equalization for single-carrier broadband wireless systems. *IEEE Communications Magazine* **40**(4), 58–66.

Fazel, K. (1993) Performance of CDMA/OFDM for mobile communication system. *IEEE Proceedings of International Conference on Universal Personal Communications*, vol. 2, pp. 975–979.

Fazel, K., Kaiser, S. and Schnell, M. (1995) A flexible and high performance cellular mobile communications system based on orthogonal multi-carrier SSMA. *Kluwer Journal of Wireless Personal Communications* **2**(1/2), 121–144.

Foschini, G. J. (1996) Layered space–time architecture for wireless communication in a fading environment when using multiple antennas. *Bell Laboratories Technical Journal* **1**(2), 41–59.

Foschini, G. and Gans, M. (1998) On limits of wireless communications in a fading environment when using multiple antennas. *Kluwer Journal of Wireless Personal Communications* **6**(3), 311–335.

Gesbert, D., Boelcskei, H., Gore, D. and Paulraj, A. (2002) Outdoor MIMO wireless channels: Models and performance prediction. *IEEE Transactions on Communications* **50**(12), 1926–1934.

Harashima, H. and Mikayawa, H. (1972) Matched-transmission technique for channels with intersymbol interference. *IEEE Transactions on Communications* **20**(08), 774–780.

Hochwald, B. and ten Brink, S. (2003) Achieving near-capacity on a multiple-antenna channel. *IEEE Transactions on Communications* **51**(03), 389–399.

IEEE (2005) IEEE 802.16e, physical and medium access control layers for combined fixed and mobile operation in licensed bands.

IEEE channel model (2004) IEEE 802.11-03/940-r4, channel model for high throughput wireless LANs.

Kaiser, S. (2002) OFDM code-division multiplexing in fading channels. *IEEE Transactions on Communications* **50**(8), 1266–1273.

Khaled, N., Modal, B., Heath, R., Leeus, G. and Petre, F. (2005) Quantized multimode precoding for spatial multiplexing MIMO-OFDM systems. *IEEE Proceedings of Vehicular Technology Conference*, vol. 2, pp. 867–871.

Klein, A., Kaleh, G. K. and Baier, P. W. (1996) Zero forcing and minimum mean square error equalization for multiuser detection in code division multiple access channels. *IEEE Transactions on Vehicular Technology* **45**(2), 276–287.

Lindskog, E. and Paulraj, A. (2000) A transmit diversity scheme for channels with intersymbol interference. *IEEE Proceedings of International Conference on Communications*, pp. 307–311.

Liu, Z., Xin, Y. and Giannakis, G. (2003) Linear constellation precoding for OFDM with maximum multipath diversity and coding gains. *IEEE Transactions on Communications* **51**(3), 416–427.

Louveaux, J., Vandendorpe, L. and Sartenaer, T. (2003) Cyclic prefixed single carrier and multicarrier transmission: Bit rate comparison. *IEEE Communications Letters* **7**(4), 180–182.

Love, D. and Heath, R. (2005) Multimode precoding for MIMO wireless systems. *Signal Processing* **53**(10), 3674–3687.

Ojanperä, T. and Prasad, R. (1998) *Wideband CDMA for Third Generation Mobile Communications*. Artech House Publishers.

Paulraj, A. and Kailath, T. (1994) Increasing capacity in wireless broadcast systems using distributed transmission/directional reception (DTDR). US Patent 5345599, Stanford University.

Raleigh, G. and Cioffi, J. (1998) Spatio-temporal coding for wireless communications. *IEEE Transactions on Communications* **46**(3), 357–366.

Sari, H., Karam, G. and Jeanclaude, I. (1995) Transmission techniques for digital terrestrial TV broadcasting. *IEEE Communications Magazine* **33**(2), 100–109.

Scaglione, A., Barbarossa, S. and Giannakis, G. (1999) Filterbank transceivers optimizing information rate in block transmissions over dispersive channels. *IEEE Transactions on Information Theory* **45**, 1998–2006.

Scaglione, A., Stoica, P., Barbarossa, S., Giannakis, G. and Sampath, H. (2002) Optimal designs for space–time linear precoders and decoders. *Signal Processing* **50**(5), 1051–1064.

Tarokh, V., Jafarkhani, H. and Calderbank, A. (1999) Space–time block codes from orthogonal designs. *IEEE Transactions on Information Theory* **45**(5), 1456–1467.

Tarokh, V., Seshadri, N. and Calderbank, A. (1998) Space–time codes for high data rate wireless communication: Performance criterion and code construction. *IEEE Transactions on Information Theory* **44**(2), 744–765.

Thoen, S., Van der Perre, L., Engels, M. and De Man, H. (2002) Adaptive loading for OFDM/SDMA-based wireless networks. *IEEE Transactions on Communications*.

van Nee, R., Awater, G., Morikura, M., Takanashi, H., Webster, M. and Halford, K. (1999) New high-rate wireless LAN standards. *IEEE Communications Magazine* **37**(12), 82–88.

Vandenameele, P., Van der Perre, L., Engels, M., Gyselinckx, B. and De Man, H. (2000) A combined OFDM/SDMA approach. *IEEE Journal on Selected Areas in Communications*.

Vollmer, M., Haardt, M. and Gotze, J. (2001) Comparative study of joint detection techniques for TD-CDMA based mobile radio systems. *IEEE Journal on Selected Areas in Communications* **19**(8), 1461–1475.

Wang, Z. and Giannakis, G. (2000) Wireless multicarrier communications: Where Fourier meets Shannon. *IEEE Signal Processing Magazine* **17**(3), 29–48.

Wang, Z. and Giannakis, G. (2003) Complex-field coding for OFDM over fading wireless channels. *IEEE Transactions on Information Theory* **49**(3), 707–720.

Yee, N., Linnartz, J. P. and Fettweis, G. (1993) Multicarrier CDMA in indoor wireless radio networks. *IEEE Proceedings of Personal, Indoor and Mobile Radio Communications*, vol. 1, pp. 109–113.

Zhou, S. and Giannakis, G. (2003) Single-carrier space–time block-coded transmissions over frequency-selective fading channels. *IEEE Transactions on Information Theory* **49**(1), 164–179.

3

Real Life Front-Ends

The physical layer implementation of almost any modern wireless device consists of three main parts (Figure 3.1): the antenna, the analog part of the transceiver and the digital part of the transceiver. In this chapter, we focus on the analog part of the transceiver, which we will often refer to as the 'front-end'. We want to describe the main front-end architectures used in modern designs and their main functional blocks. The aim of this description is to highlight the deviations from the ideal behavior of these blocks and to describe this deviation by an equivalent lowpass model, which will be complex when necessary. We will include in this analysis the analog-to-digital and digital-to-analog converters that are actually at the interface between the analog and the digital sections. Note that we will restrict this analysis to bandpass signals, i.e. systems where:

- the modulation/demodulation is performed digitally, although many concepts also apply to analog modulation/demodulation;
- the signal is modulated on a high frequency carrier and the modulation bandwidth is significantly smaller than the carrier frequency, such that complex modulation is possible.

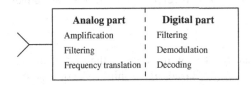

Figure 3.1 Simple partitioning of a wireless device (PHY part only).

We will first describe the main radio frequency (RF) architectures in Section 3.1. We will classify them according to the number of frequency conversions to translate the signal from RF to baseband (super-heterodyne versus direct conversion) and according to how the quadrature signals are generated (analog or digital quadrature generation). A mathematical model of the conversion from baseband to RF and vice versa will support this section. Section 3.2 will delve into the constituent blocks of the transceiver architectures. We will concentrate on amplifiers, mixers and analog-to-digital and digital-to-analog converters because these constituent blocks of the transceivers are those responsible for the main non-idealities addressed in this book. Finally, Section 3.3 will describe individually the non-idealities and provide mathematical models that we will need in the subsequent chapters of the book.

3.1 Front-End architectures

3.1.1 Mathematical model of the ideal transmitter and receiver

Ideal transmitter

We assume that a deterministic, continuous-time, complex lowpass filtered signal $x_L(t) = x_I(t) + jx_Q(t)$ is used to modulate an RF carrier at frequency f_0. The function of the ideal transmitter is to generate the bandpass signal $x_{RF}(t)$ as given by Meyr et al. (1998):

$$x_{RF}(t) = \Re[x_L(t)\, e^{j\omega_0 t}] \qquad (3.1)$$

$$= (x_I(t)\cos(2\pi f_0 t) - x_Q(t)\sin(2\pi f_0 t)) \qquad (3.2)$$

$$= \tfrac{1}{2}[(x_I(t) + jx_Q(t))\, e^{j\omega_0 t} + (x_I(t) - jx_Q(t))\, e^{-j\omega_0 t}] \qquad (3.3)$$

where $x_I(t)$ and $x_Q(t)$ are the in-phase and quadrature components of the complex baseband signal $x_L(t)$ and $\omega_0 = 2\pi f_0$.[1] Note that, although $x_L(t)$ is complex, $x_{RF}(t)$ is real. Nevertheless, it contains all the information of the complex signal $x_L(t)$; this is possible because the bandwidth of $x_{RF}(t)$ is double that of the lowpass signals $x_I(t)$ and $x_Q(t)$. The second form in Equation (3.3) is useful since it lends itself to a simple implementation (Figure 3.2). The third form in Equation (3.3) highlights the fact that the real modulated signal $x_{RF}(t)$ has both positive and negative frequency components, the carrier at $-f_0$ being modulated with the complex conjugate of $x_L(t)$, justifying the mirroring of the modulation spectrum. In Equation (3.1), $x_L(t)$ is referred to as the complex envelope of $x_{RF}(t)$.

Ideal receiver

The function of the ideal receiver is the dual of the one of the ideal transmitter, namely regenerating $x_L(t)$ from $x_{RF}(t)$. Based on the third form of Equation (3.3), it is easy to see that this can be achieved by multiplying $x_{RF}(t)$ with $e^{-j\omega_0 t}$ and rejecting the term at frequency $-2f_0$ (Figure 3.3):

$$x_{RF}(t)\, e^{-j\omega_0 t} = x_L(t)(1 + e^{-j2\omega_0 t}) \qquad (3.4)$$

Note that, if $e^{-j\omega_0 t}$ is used in Equation (3.4) instead of $e^{j\omega_0 t}$, this is equivalent to swapping the two real local oscillators (LOs) in Figure 3.3, resulting in a mirroring of the baseband spectrum. Note also that the bandwidth of $x_I(t)$ and $x_Q(t)$ is half the bandwidth of $x_{RF}(t)$.

---

[1] Although the LO in the Q branch appears with a minus sign, the reader should not interpret this as a shift towards negative frequencies. That minus sign results from the mathematical derivation in Equation (3.3), which did start with a multiplication with $e^{j\omega_0 t}$.

REAL LIFE FRONT-ENDS

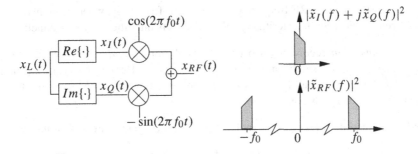

Figure 3.2 Idealized transmitter (left) and associated spectra (right).

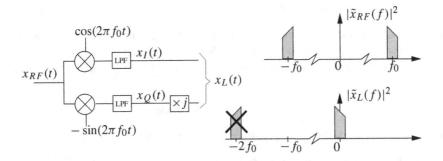

Figure 3.3 Idealized receiver (left) and associated spectra (right).

3.1.2 Classification of architectures

Although the ideal transmitter and receiver block diagrams of the previous section give insight into the operation of these functional blocks, practical systems are much more complex because of additional requirements such as linearity, selectivity and tunability. Of course, the capability of the technology also influences the front-end design. As far as the receiver is concerned, the major functions are those of frequency down-conversion, quadrature demodulation and variable gain (and the dual up-conversion and quadrature modulation apply for the transmitter).

In the following, we will use the words *band* and *channel* with the following meaning:

- a *channel* corresponds to the frequency range occupied by the desired signal ($\equiv [f_0 - f_{BW}/2; f_0 + f_{BW}/2]$),

- a *band* is a collection of neighboring channels.

A first classification of receivers is super-heterodyne versus direct conversion:

- *super-heterodyne* receivers make use of a down-conversion to baseband in at least two conversions (in other words, they make use of an intermediate frequency (IF) that is not centered on frequency 0). This usually comes with a tunable local oscillator that translates the desired RF frequency to a fixed IF frequency, where a highly selective filter rejects close-by interfering signals and selects only the desired channel;

- *direct conversion* receivers (also called *zero-IF* or *homodyne* receivers) use a LO frequency centred exactly on the channel of interest. Hence, the desired channel is translated directly to baseband where the highly selective filtering can be implemented by means of sharp lowpass filters.

A second classification of receivers is analog versus digital quadrature demodulation:

- in *analog quadrature demodulation*, the last down-conversion stage translates the channel spectrum to baseband by means of two signals in quadrature, yielding the in-phase and quadrature (I and Q) components of the complex baseband signal. Two analog-to-digital converters (ADCs) are then used to digitize the signals, with a sample rate equal to or greater than f_{BW};

- in *digital quadrature demodulation*, the desired spectrum is centered on a (usually low) IF and a single ADC is used to sample the real signal with a sample rate equal to or greater than $2 f_{BW}$. This is sometimes also referred to as *bandpass sampling* or *IF sampling*.

These system choices combine into the following alternatives:

- super-heterodyne with analog quadrature demodulation;
- super-heterodyne with digital quadrature demodulation (digital IF);
- direct conversion.

Clearly, the direct conversion receiver can only be of the 'analog quadrature demodulation' type. Other receiver architectures exist that are variant of the ones presented above. They are described in (Razavi 1998). A hybrid of the zero-IF and digital IF consists in down-converting the desired band one or two channels away from frequency 0. This technique is called 'low-IF'. It avoids some problems of the direct conversion architecture. It can be used if the adjacent channel level is not too high with respect to the desired channel. We will now describe the main receiver architectures in more detail.

3.1.3 Super-heterodyne architecture with analog quadrature

The block diagram of the super-heterodyne architecture with analog quadrature is shown in Figure 3.4. In the receive path, bandpass filters are used at RF to pre-select a frequency band of interest (typically a few channels). A low-noise amplifier (LNA) provides gain and fixes the noise floor of the system. A first conversion with a tunable LO (usually, but not always, a down-conversion) brings the signal to a fixed IF. Hence, the selectivity of the RF filter must be such that the RF image frequency, situated at $f_{RF} - 2 f_{IF}$ if $f_{LO} < f_{RF}$ or at $f_{RF} + 2 f_{IF}$ otherwise, is sufficiently attenuated.[2] At IF, a highly selective bandpass filter selects the channel of interest and attenuates all other frequencies. It is this fixed filter that

---

[2] Note that when $f_{LO} > f_{RF}$ and $f_{IF} = f_{LO} - f_{RF}$, the spectrum of the desired signal at IF is mirrored with respect to the RF spectrum.

REAL LIFE FRONT-ENDS

achieves the high selectivity in this architecture. An automatic gain control (AGC) amplifier brings the signal to a desired level for optimum analog-to-digital conversion at baseband. After the AGC amplifier, the signal is split into two equal components and down-converted to baseband with two LO signals in quadrature. The baseband signals are then lowpass filtered, to avoid any aliasing, and converted to digital. The sample rate must be at least twice as high as half the signal bandwidth: $f_S \geq f_{BW}$. The operation of the transmit part is the dual of that of the receiver.

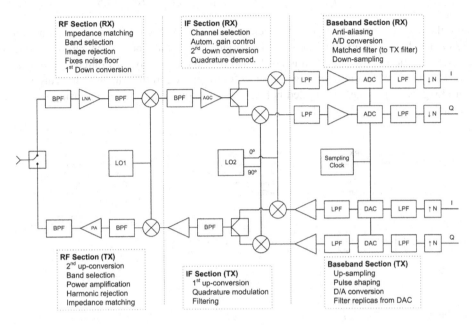

Figure 3.4 Super-heterodyne architecture with analog quadrature.

Analog quadrature generation relies on equal amplitude in the I and Q branches and exact 90° phase difference between the two LO components. Departure from this ideality is termed *IQ-imbalance* and requires special treatment; it will be considered later in this and other chapters.

3.1.4 Super-heterodyne architecture with digital quadrature

The block diagram of the super-heterodyne architecture with digital quadrature is shown in Figure 3.5. Its operation is similar to the previous architecture up to the output of the AGC amplifier. There, the real passband signal is directly sampled at IF, resulting in a real signal at a center frequency f_{IF2}. For Nyquist sampling, the sample rate must be at least twice as high as the IF frequency augmented with half the signal bandwidth: $f_S \geq 2(f_{IF} + f_{BW}/2) = 2f_{IF} + f_{BW}$. The resulting signal is then digitally processed by the following operations: complex mixing to (complex) baseband, lowpass filtering and decimation yielding the digital I and Q branches. The operation of the transmit part is the dual of that of the receiver. An interesting variant consists in sampling the IF signal below the Nyquist frequency (sub-sampling). This is possible if the analog bandwidth of the ADC extends at least to $f_{IF} + f_{BW}/2$ (although $f_S \ll f_{IF}$). It is easy to derive that the center frequency f_{IF2} of the signal after sub-sampling is given by:

$$f_{IF2} = f_{IF} - \lfloor f_{IF}/f_S \rfloor \cdot f_S \quad \text{if} \quad \text{frac}(f_{IF}/f_S) < 0.5 \quad (3.5)$$

$$= \lceil f_{IF}/f_S \rceil \cdot f_S - f_{IF} \quad \text{if} \quad \text{frac}(f_{IF}/f_S) > 0.5 \quad (3.6)$$

where the operator frac(x) denotes the decimal part of x. Sub-sampling performs implicitly a frequency down-conversion. The sampling frequency f_S must be chosen carefully so that the useful portion of the spectrum is not folded over the ADC Nyquist range (see Vaughan et al. (1991) for a detailed coverage of sub-sampling). A careful bandpass filtering of the desired signal at f_{IF} must be applied before sub-sampling because of the aliasing (of noise or signals) implicit in sub-sampling.

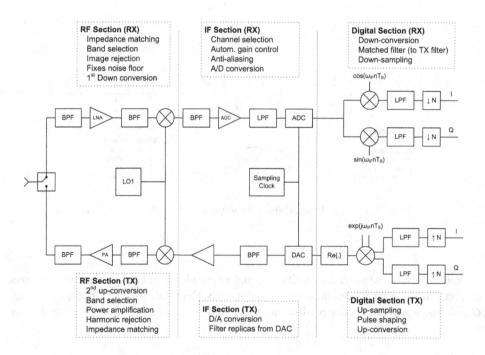

Figure 3.5 Super-heterodyne architecture with digital quadrature.

In both the Nyquist sampling and sub-sampling approaches, if f_{IF2} is chosen to be exactly $f_S/4$, the digital complex mixing is simplified into mixing with repeated sequences of 1, 0, −1, 0 or 0, 1, 0, −1, which are trivial digital operations. Sub-sampling cannot be used as easily in the transmit path because the staircase output of the DAC results in a sinc envelope in the frequency domain with zeros at integer multiple of f_S. Hence, the output amplitude of the spectral replicas are severely attenuated in the higher-order sidelobes, which prevents the use of high sub-sampling ratio in the transmitter.

3.1.5 Direct conversion architecture

The block diagram of the direct conversion architecture is illustrated in Figure 3.6. In the receive path, bandpass filters are used at RF to pre-select a frequency range of interest (typically a few channels). An LNA provides gain and fixes the noise floor of the system. An AGC amplifier brings the signal to a desired level for optimum analog-to-digital conversion at baseband. After the AGC amplifier, the signal is split into two equal components and directly down-converted to baseband with two LO signals in quadrature. The baseband signals are then lowpass filtered for two purposes: to avoid any aliasing and to reject adjacent channels. The next operation is analog-to-digital conversion. The sample rate must be at least twice as high as half the signal bandwidth: $f_S \geq f_{BW}$. The operation of the transmit part is the dual of that of the receiver. In the direct conversion receiver, selectivity is achieved by the lowpass filters at baseband.

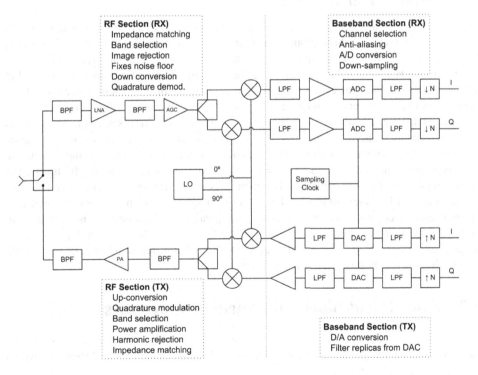

Figure 3.6 Direct conversion and Low-IF architecture.

Although the analog quadrature generation block diagram of the super-heterodyne with analog quadrature and the direct conversion architectures look similar, there is a significant difference between them: in the direct conversion receiver, a band is translated to baseband; hence, the amplifier and mixer dynamic range must be much higher and it is the LPF before the ADC that plays the role of channel filter.

Besides IQ imbalance, direct conversion receivers also suffer from DC offsets (see Section 3.3.6).

3.1.6 Low-IF architecture

The block diagram of the Low-IF receiver is similar to that of the direct conversion receiver (Figure 3.6). The main difference lies in the fact that, in the Low-IF receiver, the desired channel is translated near DC, one or two channels away from DC, in order to avoid the DC offset problem. This is only possible for standards where the adjacent channel interference levels are guaranteed to be low in the nearby channels and gradually increase for interferences further away from the desired channel. Clearly, the Low-IF architecture is not suitable for wideband modulations since the resulting sample rate would become prohibitive.

3.1.7 MIMO FE architecture

MIMO techniques are becoming increasingly important in emerging wireless standards thanks to the increase in throughput or link robustness provided by the spatial dimension. MIMO techniques rely on multiple antennas and transceivers. The individual transceivers can be of the types described above (super-heterodyne, direct conversion or low-IF) but the least complex and power-consuming architectures are often preferred because of the multiplicity of transceivers. Although the individual transceivers of a MIMO device could be implemented independently, it is better to share the LO and the ADC sampling clock. In this way, the estimation and compensation of the synchronization offsets and the compensation of phase noise is common to all receive branches, resulting in a simplified implementation and better performances (note that this sharing is also mandatory at the MIMO transmitter side). On the other hand, because MIMO techniques are usually exploited in space-selective environments, independent AGC must be used in each antenna branch to cope with significant differences in received power per antenna. A typical MIMO transceiver is illustrated in Figure 3.7 for a direct conversion architecture. See Section 6.1.3 for more in-depth coverage of AGC and synchronization in MIMO receivers. Note that different IQ imbalances in each antenna branch will require a special treatment.

3.2 Constituent blocks and their non-idealities

This section is very short. Its aim is to introduce the non-idealities originating in amplifiers, the mixing process and A/D and D/A conversion. These non-idealities will then be covered in detail in the next section.

3.2.1 Amplifiers

Amplifiers are very important in transmitters and receivers; they bring the desired signal to a suitable level for transmission or for demodulation. The two main impairments due to non-ideal amplifiers are:

REAL LIFE FRONT-ENDS

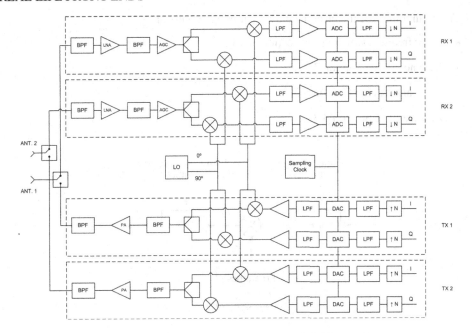

Figure 3.7 MIMO architecture.

- nonlinearities. These are characterized by the 1 dB compression point and the third-order intercept point, introduced in detail in Section 3.3.1;
- additive (thermal) noise. Additive white Gaussian noise (AWGN) is of major concern is most wireless demodulators because it determines the sensitivity of the receiver. We will describe the source and modeling of AWGN in Section 3.3.2.

3.2.2 Mixers and local oscillators

Basic operation

The function of a down- or up-conversion mixer is to translate the signal in the frequency domain, ideally without altering its characteristics. This is usually performed by multiplying the signal with a 'local oscillator', which is ideally a pure single frequency sine wave. We will discuss the down-conversion only (up-conversion is a dual operation). The signal at frequency f_{RF} is translated, by multiplication with a signal at frequency f_{LO}, to frequency $f_{IF} = f_{RF} - f_{LO}$ (Figure 3.8). This is the desired output. There is also, however, an output at frequency $f_{RF} + f_{LO}$ that must be suppressed *after* the mixer, which is easy since this term is usually at very high frequency with respect to the IF frequency. At RF, the image frequency, located at $f_{RF} - 2f_{IF}$ must be suppressed *before* the mixer. Otherwise, any signal at this frequency will also be converted to f_{IF} and will severely distort the desired signal. Note that mixers also exhibit some nonlinearity and are characterized by their gain, 1 dB compression point and third-order intercept point, as with amplifiers.

46   DIGITAL COMPENSATION FOR ANALOG FRONT-ENDS

The main impairments introduced by the mixing process are:

- carrier frequency offset (see Section 3.3.3);
- phase noise (see Section 3.3.4);
- IQ imbalance (see Section 3.3.5);
- DC offset (see Section 3.3.6).

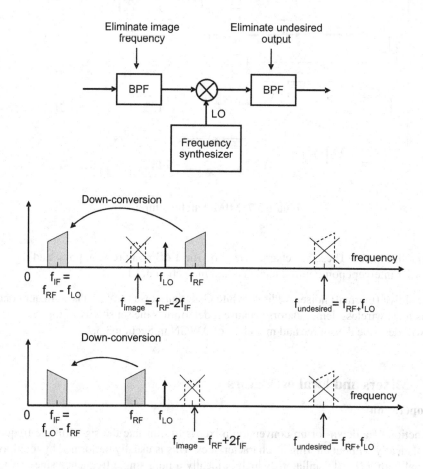

Figure 3.8 Basic mixer and LO.

3.2.3 A/D and D/A converters

A/D and D/A converters are at the heart of any modern modem. They allow powerful signal processing at the TX and RX side to interface with the analog front-ends. The process of converting an analog signal to digital (or vice versa) brings along several non-idealities:

- quantization noise (see Section 3.3.7);
- clipping (see Section 3.3.7);
- sampling clock offset (see Section 3.3.8);
- sampling jitter (see Section 3.3.9).

3.3 Individual non-idealities

This section introduces each non-ideality in detail. For most non-idealities, a Matlab model is provided in Appendix C.

3.3.1 Nonlinear amplifiers

This section deals with nonlinearities generated by amplifiers. Although any amplifier can be driven into saturation, the power amplifier in the transmitter is the most frequent source of nonlinearity. For obvious reasons of power consumption, the current in the amplifier is limited as much as possible whereas the output power is desired to be as high as possible.

RF characteristics

The conventional RF parameters used to describe an amplifier are its power gain, output-referred 1 dB compression point (OP1), saturated output power (P_{SAT}) and input-referred third-order intercept point (IIP3). These parameters are illustrated in Figure 3.9. Note that OP1 and IIP3 are power values. OP1 is defined as the output level at which the gain has reduced by 1 dB with respect to its linear characteristic. This is usually somewhat lower than the saturated output power, which is the output power where the PA is driven completely in saturation. IIP3 is a fictitious point in the input–output power characteristic of a PA. If two equal power tones at frequency f_1 and f_2 were fed simultaneously at the input of a PA, intermodulation terms at frequencies $nf_1 \pm mf_2$ would result. Among these terms, the most disturbing ones are the third-order terms of the form $2f_1 \pm f_2$ or $2f_2 \pm f_1$ because, if f_1 and f_2 are within the system bandwidth, the third-order terms fall into or close to the system bandwidth as well (Figure 3.10); the terms at $3f_1$ and $3f_2$ are usually outside the system bandwidth and are easily filtered out. The 'third order gain' (i.e. the ratio between the signal power at f_1 or f_2 and the power of the third-order term at $2f_1 \pm f_2$ or $2f_2 \pm f_1$) is also shown in Figure 3.9. The intersection between the linear extensions of the gain curve and third-order curve is the third-order intercept point. The dotted lines indicate that this is not a practical operating point for the amplifier; it is, however, widely used by RF designers as a figure of merit to describe the linearity of amplifiers.

Analytical nonlinear models

The nonlinear behavior of an amplifier can be described in a general manner by a power series expansion as follows:

$$y(t) = \sum_{k=1}^{K} a_k x_{RF}^k(t) \tag{3.7}$$

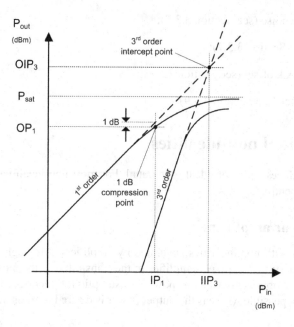

Figure 3.9 Amplifier gain and third-order curves, illustrating the 1 dB compression point and the third-order intercept point.

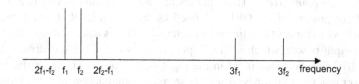

Figure 3.10 Two-tones third-order intermodulation—spectral components.

Given Equation (3.3), the kth power of $x_{RF}(t)$ is equal to:

$$x_{RF}^k(t) = \frac{1}{2^k}[x_L(t)\,e^{j\omega_0 t} + x_L^*(t)\,e^{-j\omega_0 t}]^k \tag{3.8}$$

$$= \frac{1}{2^k}\sum_{p=0}^{k}\binom{k}{p}x_L^p(t)(x_L^*(t))^{k-p}\,e^{(j\omega_0 t)(-k+2p)} \tag{3.9}$$

Assuming narrowband modulation ($f_{BW} \ll f_{RF}$), we are only interested in those terms of Equation (3.9) close to frequency f_0, i.e. those for which $-k+2p = \pm 1$, which also implies that k must be odd. Therefore, taking $k = 2m+1$, Equation (3.9) simplifies into:

$$x_{RF}^k(t) = \frac{1}{2^k}\binom{2m+1}{m+1}|x_L(t)|^{2m}[x_L(t)\,e^{j\omega_0 t} + x_L^*(t)\,e^{-j\omega_0 t}] \tag{3.10}$$

REAL LIFE FRONT-ENDS

and the complete signal at RF affected by nonlinearity is given by:

$$y(t) = \sum_{k=1}^{K} a_k x_{RF}^k(t) = \sum_{m=0}^{(K-1)/2} a_{2m+1} x_{RF}^{2m+1}(t) \qquad (3.11)$$

$$= \sum_{m=0}^{(K-1)/2} \frac{a_{2m+1}}{2^{2m+1}} \binom{2m+1}{m+1} |x_L(t)|^{2m} [x_L(t) \, e^{j\omega_0 t} + x_L^*(t) \, e^{-j\omega_0 t}] \qquad (3.12)$$

Using Equation (3.3), the lowpass complex equivalent of the nonlinearity is:

$$y_L(t) = x_L(t) \sum_{m=0}^{(K-1)/2} \frac{a_{2m+1}}{2^{2m}} \binom{2m+1}{m+1} |x_L(t)|^{2m} \qquad (3.13)$$

Equation (3.13) tells us how nonlinear coefficients a_i translate into a nonlinear model at baseband. For a third-order nonlinearity, we have $K = 3$, hence $m \in \{0, 1\}$ and Equation (3.13) reduces to

$$y_L(t) = x_L(t) \left[a_1 + \frac{3a_3}{4} |x_L(t)|^2 \right] \qquad (3.14)$$

as in Razavi (1998). Note that a_3 must be negative to result in a saturating behavior. For the case of a real third-order linearity, it is possible to relate analytically the RF parameters of the previous sub-section with the third-order term a_3 in Equation (3.14) as indicated in Table 3.1.[3] In a third-order model, power values of $IIP3$ and $IP_{1\,dB}$ are linked as follows: $IIP3 - IP_{1\,dB} = 9.6$ dB. A lowpass equivalent of a third-order amplitude transfer function is illustrated in Figure 3.11. By nature, the third-order model is non-monotonic, with the magnitude of the ouput voltage dropping as $|v_{IN}|$ is increased beyond the saturation point. Very often, this is prevented in simulation by holding the output voltage to the saturation voltage for input values exceeding the input saturation voltage (dash-dotted line in Figure 3.11).

Table 3.1 Relation between RF parameters and nonlinear model parameters (third-order only). All values starting with A represent peak amplitude values (as opposed to values in dB).

	Input-referred value	Output-referred value
1 dB compression point	$AI_{1dB} = \sqrt{\frac{(1-10^{-1/20})4a_1}{3|a_3|}}$	$AO_{1dB} = AI_{1dB} a_1 (10^{-1/20})$
Third-order intercept	$AIIP3 = \sqrt{\frac{4a_1}{3|a_3|}}$	$AOIP3 = a_1 AIIP3$
Saturation	$AI_{sat} = \frac{2}{3}\sqrt{\frac{a_1}{3|a_3|}}$	$AO_{sat} = \frac{4a_1}{9}\sqrt{\frac{a_1}{3|a_3|}}$

---

[3] a_1 can be interpreted as the voltage gain or square root of the power gain. Power values are usually given in dBmilliwatt (dBm); assuming a reference impedance of Z_0 ohm and a sinusoidal signal, root-mean-square (RMS) voltage values are derived from values in dBm as follows: $v_{peak} = \sqrt{2} v_{RMS} = \sqrt{2 Z_0 \cdot 10^{(P_{dBm}-30)/10}}$.

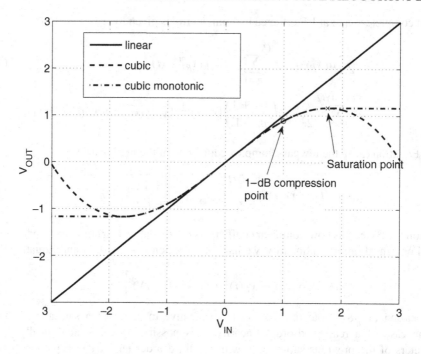

Figure 3.11 Third-order input–output function.

Ad hoc nonlinear models: AM-to-AM and AM-to-PM characteristics

In practice, the parameters of the nonlinear model in Equation (3.7) are difficult to extract from measurement, especially for the higher orders. Based on the observation that the nonlinearity in Equation (3.13) depends only on the amplitude of the signal, RF practitioners have developed two metrics that are useful for system level performance assessment: the AM-to-AM and AM-to-PM (AM–AM and AM–PM, in short) characteristics. Expressing the complex baseband signal as $x_L(t) = a_L(t) e^{j\phi(t)}$, the AM–AM characteristic $f(a_L(t))$ and AM–PM characteristic $g(a_L(t))$ describe how the instantaneous input amplitude affects the output amplitude and phase, respectively. The output of the model is then (see also Figure 3.12):

$$y_L(t) = f(a_L(t)) e^{j(\phi(t) + g(a_L(t)))} \tag{3.15}$$

Some points of these characteristics can be measured in the laboratory. From these measurements, functions are derived empirically that approximate the observed behavior. Common AM–AM and AM–PM models are the third-order model (previous sub-section), the Rapp model (Rapp 1991), the modified Rapp model and the Saleh model (Saleh 1981). These mathematical models are detailed in Table 3.2. An example of distortion on a 16QAM constellation due to AM–AM and AM–PM is provided in Figure 3.13.

REAL LIFE FRONT-ENDS

Table 3.2 Common AM–AM and AM–PM models.

	$f(a_L(t))$	$g(a_L(t))$
Third-order	$a_L(t)\left(a_1 + \dfrac{3a_3}{4}|a_l(t)|\right)$	1
Rapp	$\dfrac{G|a_L(t)|}{(1+(\frac{G|a_L(t)|}{V_{sat}})^{2p})^{\frac{1}{2p}}}$	1
Modified Rapp	$\dfrac{G|a_L(t)|}{(1+(\frac{G|a_L(t)|}{V_{sat}})^{2p})^{\frac{1}{2p}}}$	$\dfrac{A|a_L(t)|^q}{(1+(\frac{|a_L(t)|}{B})^q)}$
Saleh	$\dfrac{A|a_L(t)|}{1+B|a_L(t)|^2}$	$\dfrac{C|a_L(t)|^2}{1+D|a_L(t)|^2}$

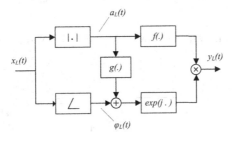

Figure 3.12 AM–AM and AM–PM functional model.

Spectral regrowth

The lowpass equivalent model of the nonlinearity is the best starting point to explain spectral regrowth. Assuming a third-order nonlinearity, the output signal has the following form:

$$y_L(t) = x_L(t) + \alpha_3 x_L^2(t) x_L^*(t) \qquad (3.16)$$

For a deterministic signal $x_L(t)$, the Fourier transform of $y_L(t)$ takes the form:

$$\tilde{y}_L(f) = \tilde{x}_L(f) + \alpha_3 \tilde{x}_L(f) \otimes \tilde{x}_L(f) \otimes \tilde{x}_L^*(-f) \qquad (3.17)$$

The convolutions in the second term of Equation (3.17) are responsible for the widening of the spectrum. It should be noted that this analysis is not totally accurate since we should actually compute the power spectral density (PSD) of $y_L t$ instead of its Fourier transform (i.e. we should compute the Fourier transform of the average auto-correlation function of $y_L(t)$). However, this is mathematically quite involved; the interested reader is referred to Gard et al. (1999) or Nsenga et al. (2008) for a detailed analysis of the PSD with nonlinearity. A numerical example of spectral regrowth, for a third-order nonlinearity as in Equation (3.14), is illustrated in Figure 3.14.

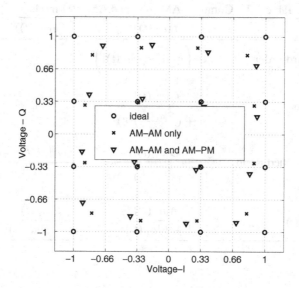

Figure 3.13 Example of the effect of AM–AM and AM–PM on a QAM16 constellation.

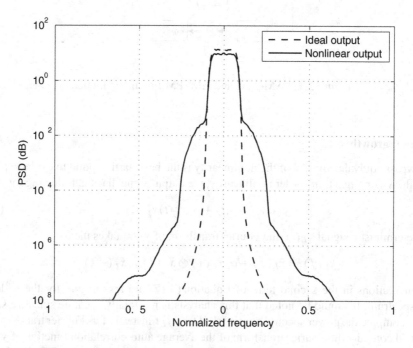

Figure 3.14 PSD of signal with ideal amplification (dashed) and nonlinear amplification (solid).

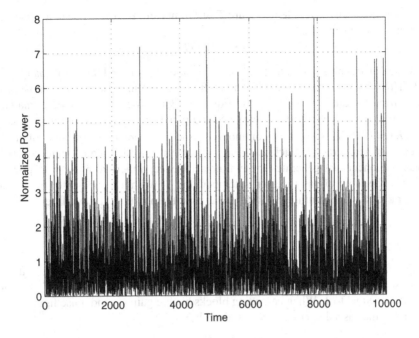

Figure 3.15 Normalized output power versus time for a 16QAM OFDM signal. Note the high peaks that are responsible for the high PAPR.

Peak-to-average power ratio

The sensitivity of a signal to nonlinearities is closely linked with its complex envelope. Some signals have a constant envelope and are not sensitive to saturation in amplifiers: this is the case for CPM. Other signals that do not enjoy the constant envelope property are severely distorted by nonlinearities, such as OFDM signals (see Figure 3.15). A classical way to characterize the envelope of a signal is the PAPR which is defined as follows:

$$PAPR = \frac{\max |x_{OFDM}(t)|^2}{\frac{1}{QT} \int_0^{QT} |x_{OFDM}(t)|^2 \, dt} \quad (3.18)$$

where $x_{OFDM}(t)$ is a time domain OFDM signal (Q is the number of sub-carriers and T is the baseband sample duration). This definition measures the PAPR over a single OFDM symbol. Usually, one has to average the PAPR over many OFDM symbols to have a better estimate of the PAPR. The crest factor is also widely used and it is defined simply as the square root of the PAPR.

3.3.2 Noise in amplifiers (AWGN)

Every transmitter and receiver needs amplifiers to bring the signal to the desired levels. Another *unavoidable* source of non-ideality due to amplifiers is the added thermal noise.

The electrons in a resistor R at temperature T different from zero give rise to a noise voltage whose mean square value is:

$$\overline{e_n^2} = 4RkTB \tag{3.19}$$

where k is Boltzmann's constant (1.38×10^{-23} joule/deg.), and B is the bandwidth of interest. This can be represented as a noise-free resistor in series with a noise voltage source with mean-square value given by Equation (3.19). When the source is matched to the load, optimum power transfer occurs and the power transferred to the load R is given by $\overline{e_n^2}/(4R) = kTB$. This noise is commonly termed *thermal noise* because of its linear dependence on temperature. Each amplifier is made of physical components including a resistive component that generates thermal noise. Hence, a thermal noise with power kTB is present at the input of any receiver at temperature T (within a bandwidth B). The *noise figure* characterizes the resulting SNR degradation at the output of the amplifier:

$$F = \frac{S_{in}/N_{in}}{S_{out}/N_{out}} = \frac{N_{out}}{kTBG} \tag{3.20}$$

where S_{in} and S_{out} are the input and output signal powers, N_{in} and N_{out} are the input and output noise powers and G is the power gain $G = S_{out}/S_{in}$. The noise figure F is obviously greater than or equal to 1. When cascading blocks having gain G_i and noise figure F_i, it can be shown that the cascaded noise figure is given by:

$$F_{tot} = F_1 + \frac{F_2 - 1}{G_1} + \frac{F_3 - 1}{G_1 G_2} + \cdots + \frac{F_N - 1}{G_1 G_2 \cdots G_{N-1}} \tag{3.21}$$

This shows that, in a properly designed receiver, it is important to have a low noise figure and a high gain in the first amplifier. An example of calculation of signal power, noise power and noise figure in a cascade is shown in Figure 3.16. Our discussion here addressed amplifiers but the gain and noise figure description is valid for any block in the receiver chain. In particular, a lossy component such as an attenuator with gain $G_{Att} = 1/L_{Att}$ has a noise

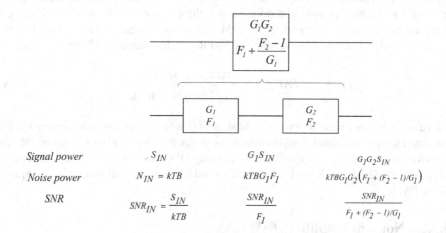

Figure 3.16 Example of cascaded noise figure calculation.

REAL LIFE FRONT-ENDS

figure $F_{Att} = L_{Att}$. A receiver input consisting of an attenuator ($G_{Att} = 1/L_{Att}$, $F_{Att} = L_{Att}$), followed by an amplifier (G_{Amp}, F_{Amp}) has, applying Equation (3.21), a gain of G_{Amp}/L_{Att} and a noise figure of $L_{Att} + F_{Amp}$: the noise figure of the attenuator adds to the noise figure of the amplifier. This is the reason why losses must be minimized before the first amplifier. The advantage of calculating the cascaded gain and noise figure of the receiver is that the complete receiver can be modelled as a single noise source with power $kTBF_{tot}$ followed by an ideal amplifier with gain G_{tot} (Figure 3.17). If the received signal power S is known, the SNR is equal to $S/(kTBF_{tot})$. For analytical or simulation models, we can then take $\sigma_n^2 = SNR^{-1} = (kTBF_{tot})/S$ for the noise variance and normalize the samples $x(n)$ at the receiver input such that their variance σ_x^2 is equal to S.

Figure 3.17 Equivalent noise model for analysis and simulations.

3.3.3 Carrier frequency offset

After the last down-conversion (which can also be done digitally), the signal is ideally a complex signal centred exactly on 0 Hz. In order to achieve this, the receiver analog and/or digital LO(s) must be set exactly at the correct frequency. Since the synthesizers of the transmitter and the receiver derive their frequency from a crystal oscillator, significant differences may exist. For example, if the crystals have an accuracy of ±20 ppm, which is typical for portable consumer electronics, and the carrier frequency is 5 GHz, a difference of $2 \times 5 \times 10^9 \times 20^{-5} = 100$ kHz may exist. This means that the received baseband signal undergoes a rotation of 2π every 10 μs, resulting in complete loss of information unless counter-measures are taken (synchronization).

Another cause of carrier frequency offset (CFO) is Doppler shift: if the transmitter and receiver are in relative motion with relative velocity v_{rel}, the received signal can be expressed as:

$$r(t) = \Re[x_L(t) \, e^{j\omega_0 t} \, e^{j\omega_{Dop} t}] = \Re[x_L(t) \, e^{j\omega_0(1+v_{rel}/c)t}] \tag{3.22}$$

where $\omega_{Dop} = 2\pi v_{rel}/\lambda$, c is the speed of light and λ is the wavelength. A speed of 160 km/h results in a ratio of $v/c = (100/3.6 \times 3 \times 10^8) \cong 10^{-7} = 0.1$ ppm. Although not negligible, CFO due to Doppler shifts are usually lower than the CFO due to crystal inaccuracies.[4]

---

[4] A notable exception is in satellite communication receivers, where speeds can be orders of magnitude higher than for terrestrial systems.

Table 3.3 Responses of PLL to different noise sources.

Reference noise	$H_{Ref}(s) = \dfrac{\theta_{out}(s)}{\theta_{Ref}(s)}$	$= \dfrac{K_{PD}K_{VCO}F(s)}{s + \dfrac{K_{PD}K_{VCO}F(s)}{N}}$
VCO noise	$H_{VCO}(s) = \dfrac{\theta_{out}(s)}{\theta_{VCO}(s)}$	$= \dfrac{s}{s + \dfrac{K_{PD}K_{VCO}\cdot F(s)}{N}}$

Modeling CFO

CFO is quite easy to model, since it is merely a frequency shift (Δf) of the complex lowpass signal:

$$y_L(t) = x_L(t)\, e^{j\Delta ft} \qquad (3.23)$$

3.3.4 Phase noise

Unfortunately, the LO is not a pure single frequency sine wave. This is due to the process by which the LO is generated in the synthesizer. We have to delve inside the architecture of a typical frequency synthesizer in order to understand the source of non-ideality. The key elements of a phase-locked loop-based frequency synthesizer comprises a crystal oscillator (the reference), a phase detector, a loop filter, a voltage-controlled oscillator and a programmable frequency divider (Figure 3.18, top). When the loop is locked, the voltage-controlled oscillator (VCO) frequency f_{VCO} is equal to the reference frequency multiplied by N: $f_{VCO} = N \cdot f_{ref}$.[5] In this way, f_{VCO} can be tuned by changing N, with a resolution equal to f_{ref}. The reason for using such an architecture is that it is very difficult to realize VCOs with a spectrally pure output: the PSD of a free-running VCO is extremely bad because the Q-factor of the VCO resonant circuit is typically below 100 whereas the Q of a crystal oscillator is of the order of 10^5 (in a resonant circuit, the 3 dB width of the resonance Δf is given by $\Delta f = f_0/Q$ where f_0 is the resonant frequency, and Q is the *quality factor*). The phase-locked loop equivalent system (Figure 3.18, bottom) allows the response of the phase-locked loop (PLL) to the phase noise coming from the reference and the VCO (θ_{Ref} and θ_{VCO}) to be derived. These responses are detailed in Table 3.3.

For a typical PLL-based frequency synthesizer, the general shape of these transfer functions is illustrated in Figure 3.19. We observe that the PLL behaves as a lowpass filter (with gain N^2 in the passband) for the reference noise and as a highpass filter (with unity gain in the passband) for the VCO noise. In other words, the output phase noise close to the carrier is dominated by the reference noise multiplied by N^2 whereas the output phase noise far from the carrier is dominated by the VCO noise. This allows a considerable improvement of the phase noise at small offsets from the carrier.

Effect of phase noise on linear modulation

The effect of phase noise is a phase modulation of the signal of the local oscillator. This phase modulation is transferred directly onto the desired signal so that the complex lowpass

---

[5] More complex architectures derived from this canonical PLL-based synthesizer exist, such as the fractional-N architecture or the direct-digital-synthesizer driven PLL that improve the resolution (Manassevitch 1962).

REAL LIFE FRONT-ENDS

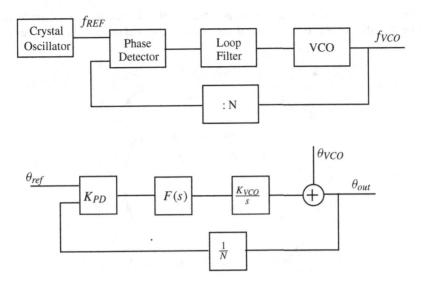

Figure 3.18 PLL block diagram (top) and linearized equivalent system diagram (bottom).

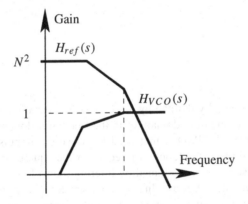

Figure 3.19 PLL transfer functions.

equivalent signal can be easily expressed as:

$$y_L(t) = x_L(t) e^{j\phi(t)} \tag{3.24}$$

where $\phi(t)$ is a time domain description of the phase noise process. Since $\phi(t)$ is usually $\ll 1$, $e^{j\phi(t)} \cong 1 + j\phi(t)$, so the effect of phase noise is approximately a perturbation in quadrature with respect to the received constellation, as illustrated in Figure 3.20. This approximation allows the modulation by the phase noise to be considered as a *narrowband phase modulation* (*PM*) whereby the PSD of the phase modulated LO is the same as that of $\phi(t)$. It is important to note that phase noise is a *multiplicative* process. In other words, increasing the SNR does not reduce the effect of phase noise but only reduces the effect of other additive noise sources.

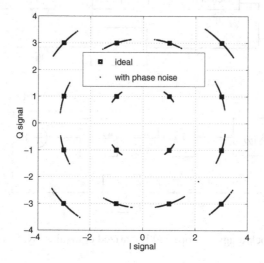

Figure 3.20 Effect of phase noise on a QAM constellation (integrated phase noise = -25 dBc).

Modeling phase noise

The phase noise PSD is usually defined by a piece-wise linear model. There are two basic approaches: generation in the time-domain or in the frequency domain. In the time domain approach (Figure 3.21(a)), the PSD is first used to define the impulse response of a filter whose frequency-domain transfer function is equal to the square root of the PSD. Then, the samples of the phase noise are obtained by filtering (convolving) a real Gaussian noise source with this filter impulse response. In the frequency domain approach (Figure 3.21(b)), a complex Gaussian noise sequence is first generated. It is then element-wise multiplied with the square root of the desired PSD and converted to the time domain by means of an inverse Fourier transform. Note that the frequency domain sequence must be complex conjugate symmetric about DC so that its corresponding time domain signal is purely real.

3.3.5 IQ imbalance

Conversion with quadrature local oscillators in the analog domain is often used in the transmitter and/or receiver sections (Section 3.1.2). This convenient implementation of quadrature conversion is affected by phase and amplitude offsets in the two branches, and is referred to as IQ imbalance. This imbalance can happen at the transmitter, receiver or both. We will treat the three cases in detail because they lead to slightly different models. Receiver IQ imbalance will be treated first.

REAL LIFE FRONT-ENDS

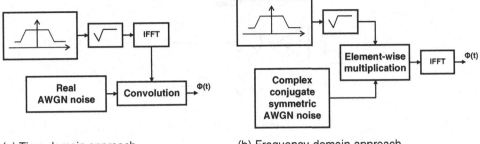

(a) Time-domain approach   (b) Frequency-domain approach

Figure 3.21 The phase noise can be generated (a) in the time-domain, or (b) frequency domain, and then transformed to the time-domain.

Receiver IQ imbalance

We assume an ideal transmitter has been used to up-convert the complex baseband signal $x_L(t) = x_I(t) + jx_Q(t)$, resulting in an ideal RF signal $x_{RF}(t)$. In the direct conversion or the super-heterodyne with analog quadrature architectures, two mixers driven by LO signals in quadrature are used to convert the real bandpass signal $x_{RF}(t)$ into a complex lowpass signal $x_{LP}(t)$, the two output branches being usually referred to as I and Q branches. If the receiver I and Q branches do not have equal gain or are not exactly in quadrature, signal distortion results. These differences can be represented by different gains ($1 + \epsilon_R$ and $1 - \epsilon_R$) and different phases ($+\Delta\phi_R$ and $-\Delta\phi_R$), as illustrated in Figure 3.22. The distorted representation of the signal is then, using ϵ_R and $\Delta\phi_R$ for the gain and phase imbalances and assuming no CFO and no noise:

$$x_{LP}(t) = x_{RF}(t)x_{LO}(t) \tag{3.25}$$

$$= (x_I(t)\cos(\omega_0 t) - jx_Q(t)\sin(\omega_0 t))$$
$$\times ((1+\epsilon_R)\cos(\omega_0 t + \Delta\phi_R) - j(1-\epsilon_R)\sin(\omega_0 t - \Delta\phi_R)) \tag{3.26}$$

$$= (\cos(\Delta\phi_R) - j\epsilon_R \sin(\Delta\phi_R))(x_I(t) + jx_Q(t))$$
$$+ (\epsilon_R \cos(\Delta\phi_R) + j\sin(\Delta\phi_R))(x_I(t) - jx_Q(t)) \tag{3.27}$$

$$= \alpha_R x_L(t) + \beta_R x_L^*(t) \tag{3.28}$$

in which:

$$\alpha_R = \cos(\Delta\phi_R) - j\epsilon_R \sin(\Delta\phi_R) \tag{3.29}$$

$$\beta_R = \epsilon_R \cos(\Delta\phi_R) + j\sin(\Delta\phi_R) \tag{3.30}$$

and $x_{LO}(t)$ captures both the phase and amplitude imbalances. Equation (3.28) reveals that the resulting lowpass signal $x_{LP}(t)$ consists of the ideal signal $x_L(t)$ plus a scaled version of its complex conjugate $x_L^*(t)$. Because of the complex conjugation, the positive part of the spectrum of $\tilde{x}_L(f)$ interferes with the negative part of its spectrum and vice versa.

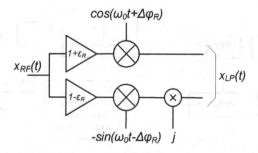

Figure 3.22 Receiver IQ imbalance equivalent diagram (amplitude offset: ϵ_R; phase offset: $\Delta\phi_R$).

The relative amount of interference (in dB) resulting from IQ imbalance is easily found as:

$$10\log\left(\frac{|\beta_R|^2}{|\alpha_R|^2}\right) = 10\log\left(\frac{\epsilon_R^2 + \tan^2(\Delta\phi_R)}{1 + \epsilon_R^2 \tan^2(\Delta\phi_R)}\right) \qquad (3.31)$$

Figure 3.23 shows, in a amplitude error versus phase error plane, lines of constant levels of interference. We see that, in order to have a signal-to-distortion-ratio of at least 30 dB, we need an amplitude mismatch smaller than 3% or a phase mismatch smaller than 2°.

Frequency-dependent IQ imbalance

With the increasing bandwidth of emerging air interfaces, the analog lowpass filters in the I and Q branches can be very broadband (several MHz, up to 1 or 2 GHz) and it becomes increasingly difficult to match them. It results in the IQ imbalance becoming frequency-dependent. The treatment is similar to that of the previous section, except that the offsets ϵ and $\Delta\phi$ are now functions of f: $\epsilon_R(f)$ and $\Delta\phi_R(f)$ or $\epsilon_T(f)$ and $\Delta\phi_T(f)$. This will clearly increase the difficulty of estimating and compensating the IQ imbalance. This will be further discussed in Section 5.10.

Transmitter IQ imbalance

The derivation of the effect of transmitter IQ imbalance is as follows. The ideal complex baseband signal $x_L(t) = x_I(t) + jx_Q(t)$ is up-converted with quadrature local oscillators affected by amplitude and phase imbalances characterized by the offsets ϵ_T and $\Delta\phi_T$ (Figure 3.24). The transmitted signal at RF can then be expressed as:

$$x_{RF}(t) = (1 + \epsilon_T)\cos(\omega_0 t + \Delta\phi_T)x_I(t) - (1 - \epsilon_T)\sin(\omega_0 t - \Delta\phi_T)x_Q(t) \qquad (3.32)$$

Assuming an ideal receiver, the receiver complex baseband signal results from mixing with $x_{LO}(t) = \exp(-j\omega_0 t)$:

$$x_{LP}(t) = x_{RF}(t)x_{LO}(t) \qquad (3.33)$$

$$= (1 + \epsilon_T)\cos(\omega_0 t + \Delta\phi_T)x_I(t)(\cos(\omega_0 t) - j\sin(\omega_0 t))$$
$$- (1 - \epsilon_T)\sin(\omega_0 t - \Delta\phi_T)x_Q(t)(\cos(\omega_0 t) - j\sin(\omega_0 t)) \qquad (3.34)$$

REAL LIFE FRONT-ENDS

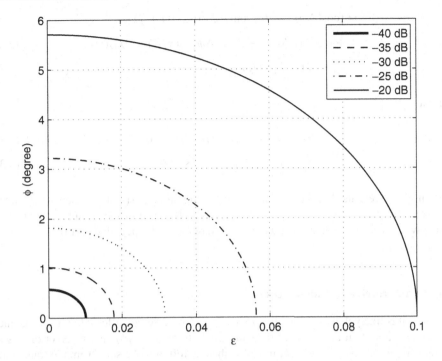

Figure 3.23 Contour of constant interference due to IQ mismatch in the amplitude error versus phase error plane. These contour plots are valid for both TX and RX IQ imbalances.

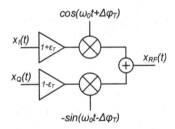

Figure 3.24 Transmitter IQ imbalance equivalent diagram (amplitude offset: ϵ_T; phase offset: $\Delta\phi_T$).

After some algebraic and trigonometric manipulations and ignoring the terms at frequency $2\omega_0 t$ that are eliminated by lowpass filtering, this is equal to:

$$x_{LP}(t) = (1 + \epsilon_T)(\cos(\Delta\phi_T) + j\sin(\Delta\phi_T))x_I(t)$$
$$+ (1 - \epsilon_T)(\sin(\Delta\phi_T) + j\cos(\Delta\phi_T))x_Q(t) \quad (3.35)$$

$$= (\cos(\Delta\phi_T) + j\epsilon_T \sin(\Delta\phi_T))(x_I(t) + jx_Q(t))$$
$$+ (\epsilon_T \cos(\Delta\phi_T) + j \sin(\Delta\phi_T))(x_I(t) - jx_Q(t)) \quad (3.36)$$
$$= \alpha_T x_L(t) + \beta_T x_L^*(t) \quad (3.37)$$

in which

$$\alpha_T = \cos(\Delta\phi_T) + j\epsilon_T \sin(\Delta\phi_T) \quad (3.38)$$
$$\beta_T = \epsilon_T \cos(\Delta\phi_T) + j \sin(\Delta\phi_T) \quad (3.39)$$

Equation (3.37) reveals that the resulting lowpass signal consists of the ideal signal $x_L(t)$ plus a scaled version of its complex conjugate $x_L^*(t)$. Because of the complex conjugation, the positive part of the spectrum of $\tilde{x}_L(f)$ interferes with the negative part of its spectrum and vice versa.

Transmit and receive IQ imbalances

In practice, it is likely that both the transmitter and the receiver are affected by IQ imbalance. We then have a system model with transmitter offsets ϵ_T and $\Delta\phi_T$ and receiver offsets ϵ_R and $\Delta\phi_R$ (Figure 3.25). The system model is then as follows. We start from the transmitted signal affected by transmit (TX) IQ imbalance as given by Equation (3.32). This signal is then down-converted with quadrature LOs also affected by receive (RX) IQ imbalances, resulting in the following receiver complex baseband signal:

$$x_{LP}(t) = ((1 + \epsilon_T)\cos(\omega_0 t + \Delta\phi_T)x_I(t) - (1 - \epsilon_T)\sin(\omega_0 t - \Delta\phi_T)x_Q(t))$$
$$\times ((1 + \epsilon_R)\cos(\omega_0 t + \Delta\phi_R) - j(1 - \epsilon_R)\sin(\omega_0 t - \Delta\phi_R)) \quad (3.40)$$

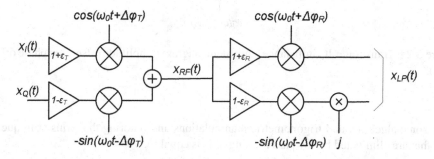

Figure 3.25 Transmitter and receiver IQ imbalance equivalent diagram (amplitude offsets: ϵ_T and ϵ_R; phase offsets: $\Delta\phi_T$ and $\Delta\phi_R$).

REAL LIFE FRONT-ENDS

After some lengthy algebraic and trigonometric manipulations and ignoring the terms at frequency $2\omega_0 t$ that are eliminated by lowpass filtering, Equation (3.40) can be expressed as:

$$x_{LP}(t) = ((1 + \epsilon_T \epsilon_R) \cos(\Delta\phi_T - \Delta\phi_R) + j(\epsilon_T - \epsilon_R) \sin(\Delta\phi_T + \Delta\phi_R))$$
$$\times (x_I(t) + jx_Q(t)) + ((\epsilon_T + \epsilon_R) \cos(\Delta\phi_T - \Delta\phi_R)$$
$$+ j(1 - \epsilon_T \epsilon_R) \sin(\Delta\phi_T + \Delta\phi_R)) \times (x_I(t) - jx_Q(t)) \quad (3.41)$$
$$= \alpha_{TR} x_L(t) + \beta_{TR} x_L^*(t) \quad (3.42)$$

in which:

$$\alpha_{TR} = (1 + \epsilon_T \epsilon_R) \cos(\Delta\phi_T - \Delta\phi_R) + j(\epsilon_T - \epsilon_R) \sin(\Delta\phi_T + \Delta\phi_R) \quad (3.43)$$
$$\beta_{TR} = (\epsilon_T + \epsilon_R) \cos(\Delta\phi_T - \Delta\phi_R) + j(1 - \epsilon_T \epsilon_R) \sin(\Delta\phi_T + \Delta\phi_R) \quad (3.44)$$

The reader can verify that Equations (3.43) and (3.44) reduce to Equations (3.29) and (3.30) when $\epsilon_T = 0$ and $\Delta\phi_T = 0$ and reduce to Equations (3.38) and (3.39) when $\epsilon_R = 0$ and $\Delta\phi_R = 0$.

Modeling IQ imbalance

IQ imbalance can be modeled directly from Equations (3.28), (3.37) or (3.42). However, the complex conjugation in these equations is not a linear operation. By splitting the complex values into real and imaginary parts, a linear model in real components only can be obtained. Using $\alpha = \alpha_r + j\alpha_i$, $\beta = \beta_r + j\beta_i$ and $x_{LP}(t) = x_{LPI}(t) + jx_{LPQ}(t)$, the IQ imbalance model can be further developed into:

$$\begin{bmatrix} x_{LPI} \\ x_{LPQ} \end{bmatrix} = \begin{bmatrix} \alpha_r + \beta_r & -\alpha_i + \beta_i \\ \alpha_i + \beta_i & \alpha_r - \beta_r \end{bmatrix} \begin{bmatrix} x_I \\ x_Q \end{bmatrix} \quad (3.45)$$

Equation (3.45) applies to the RX IQ imbalance, TX IQ imbalance or TX and RX IQ imbalance. For example, in the case of the RX IQ imbalance, Equation (3.45) becomes:

$$\begin{bmatrix} x_{LPI} \\ x_{LPQ} \end{bmatrix} = \begin{bmatrix} (1+\epsilon)\cos(\Delta\phi) & +(1+\epsilon)\sin(\Delta\phi) \\ (1-\epsilon)\sin(\Delta\phi) & (1-\epsilon)\cos(\Delta\phi) \end{bmatrix} \begin{bmatrix} x_I \\ x_Q \end{bmatrix} \quad (3.46)$$

3.3.6 DC offset

In a direct conversion receiver, the LO signal is exactly at the same frequency as the desired RF signal. Because of that, this receiver is also affected by two other non-idealities:

- DC offset: in the mixing process, part of the LO power 'leaks' to the RF port of the mixer and is mixed with itself in the mixer, resulting in two components, one at frequency 0 (DC) and one at $2 f_0$ (Figure 3.26, top). The component at DC is troublesome because it offsets the signal at the input of the analog-to-digital converter (ADC) and reduces its effective dynamic range. Hence it must be compensated for;

- slowly varying DC offset: if the LO signal leaks up to the antenna port, it can be radiated, reflected off nearby objects in relative motion, received and down-converted with the LO signal. This results in a slowly varying DC offset (Doppler effect) which makes its compensation more delicate (Figure 3.26, bottom).

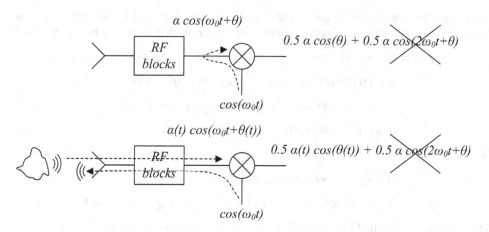

Figure 3.26 Static (top) and dynamic (bottom) DC offset in direct conversion receiver.

3.3.7 Quantization noise and clipping

The number of bits N_{bit} of the ADC and digital-to-analog converter (DAC) must be kept as low as possible for obvious reasons of cost and power consumption. On the other hand, a large number of bits is desirable to reduce the effect of quantization noise, reduce the risk of clipping and accommodate signal level variations. We start this analysis by a derivation of the SNR due to quantization. Assuming an ADC with N_{bit} bits and clipping (i.e. full-scale) levels $\pm A_{CL}$, the level of the least significant bit (LSB) or smallest analog step is $A_{LSB} = A_{CL}/2^{N_{bit}-1}$. Assuming a linear conversion slope, the amplitude error is then comprised between $-A_{LSB}/2$ and $A_{LSB}/2$. If the samples of the input signal are random, zero-mean and uniformly distributed across the quantization levels over the full-scale range of the ADC, the quantization error is uniform over $[-A_{LSB}/2, A_{LSB}/2]$. The variance of the quantization error is therefore:

$$\sigma_{Quant}^2 = \int_{-A_{LSB}/2}^{A_{LSB}/2} v^2 dv = \frac{A_{LSB}^2}{12} \qquad (3.47)$$

On the other hand, the power of a full-scale sinusoidal signal is $A_{CL}^2/2 = (2^{2N_{bit}-2} A_{LSB}^2)/2$. The SNR is therefore:

$$SNR_{ADC} = 10 \log_{10} \left(\frac{A_{CL}^2/2}{A_{LSB}^2/12} \right) = 6.02 N_{bit} + 1.76 \quad \text{(dB)} \qquad (3.48)$$

This has been derived for a sinusoidal input signal but it is representative of the best SNR to expect from an N_{bit} bits ADC. It is possible to reduce the impact of quantization noise by oversampling, i.e. sampling at a rate much higher than the Nyquist rate. Indeed, the quantization noise PSD is approximately uniformly distributed over the full Nyquist range of the ADC. Oversampling, followed by a filter selecting only the channel of interest reduces the amount of quantization noise in proportion to the ratio of the signal bandwidth f_{BW} over

REAL LIFE FRONT-ENDS

the Nyquist frequency $f_{Nyquist}$, and the SNR becomes:

$$SNR_{ADC,Oversampled} = 6.02 N_{bit} + 1.76 + 10 \log_{10}\left(\frac{f_{Nyquist}}{f_{BW}}\right) \quad \text{(dB)} \qquad (3.49)$$

Interestingly, when the signal-to-noise-and-distortion ratio (SNDR) of a given ADC is known from measurement, Equation (3.48) can be inverted to compute the effective number of bits (ENOB), which can be non-integer:[6]

$$ENOB = (SNDR - 1.76)/6.02 \quad \text{(bits)} \qquad (3.50)$$

The ENOB of an ADC depends on the input frequency (for a constant sampling frequency) in a roughly monotonic way. The effective resolution bandwidth (ERBW) of an ADC is the frequency at which the ENOB drops by half a bit with respect to the ENOB for an input at DC. Equivalently, it is the frequency at which the SNDR is degraded by 3 dB with respect to the SNDR for a DC input. Very loosely, we can express this as:

$$ENOB_{@ERBW} = ENOB_{DC} - 0.5 \quad \text{(bits)} \qquad (3.51)$$

$$SNDR_{@ERBW} = SNDR_{DC} - 3 \quad \text{(dB)} \qquad (3.52)$$

When dealing with communication signals sampled by an ADC, there are two issues to consider: first, the signal amplitude usually fluctuates according to the transmitted symbols and peaks can occur in the signal that are significantly higher than the signal RMS value; second, since the amplitude of the received signal is unknown in advance, the receiver chain must adjust its gain to make the signal fit in the ADC dynamic range. Because of the peaks in the received signal, this fitting in the ADC dynamic range is a trade-off between the amount of clipping and the amount of quantization noise: if the signal is too weakly amplified before the ADC, no clipping occurs but the ADC quantization noise is relatively high; if the signal is too strongly amplified, the impact of the quantization noise is smaller, but the signal peaks can be clipped at full-scale, causing severe distortions. Clearly, an optimum can be found. This is illustrated in Figure 3.27 that shows the resulting SNDR of a quantized 16QAM waveform. The horizontal axis shows the relative clipping level, which is the ratio between the signal RMS value and the ADC clipping level.

3.3.8 Sampling clock offset

Sampling clock offset (SCO) is somehow related to CFO but its effect and impact are different. Since all clocks and local oscillators are usually derived from a single reference oscillator (crystal oscillator), the accuracy of all these signals is equal to that of the reference (Figure 3.28). The effect of SCO is complex and depends on the modulation and down-conversion schemes. We will consider a simple linear modulation with Nyquist filtering and AWGN channel here to introduce the basic problem and refer the reader to Sections 4.1.3 and 4.2.2 for more complex schemes. Basically, if the sampling frequency $F_S = (1/T_S)$ is offset by a factor δ with respect to the ideal sampling frequency, the sampling points

---

[6] An ADC SNDR can be obtained by sampling a pure sinewave, computing the spectrogram from the samples and computing the ratio of the power in the signal spectral line over the total power in all other frequency bins.

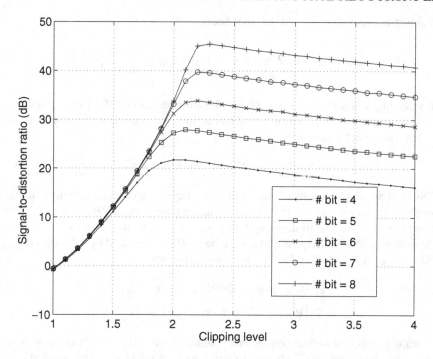

Figure 3.27 Impact of clipping level and resolution on linear modulation.

become $(1, 2, \ldots, N)(1 + \delta)T_S$ instead of $(1, 2, \ldots, N)T_S$ such that a time shift equal to $n\delta T_S$ appears on the nth sample. Several problems result (Figure 3.29). First, because of the increasing sampling time error, the signal is no longer sampled at the optimum point in the eye diagram (point of maximum opening) and degradation occurs. Second, after a time $t = T_S/\delta$, the receiver will have a sampling time shifted by one complete sample. In order to avoid these problems, it is in most cases necessary to measure and compensate the SCO.

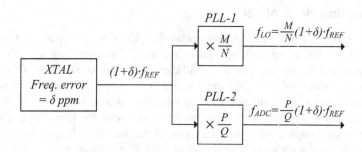

Figure 3.28 Relative error δ on reference appears on all LOs and clocks.

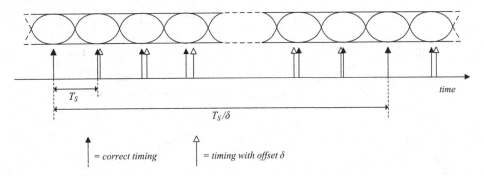

Figure 3.29 Sample point drift due to SCO and sampling position with respect to received symbols.

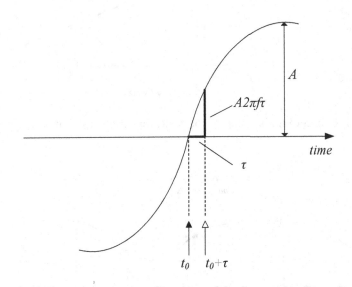

Figure 3.30 Impact of timing jitter: for a sinusoidal signal, a timing error τ has the highest impact near the zero crossing because the slope is the highest.

3.3.9 Sampling jitter

Clock jitter is another source of degradation related to digital-to-analog conversion. Due to clock jitter, the samples are taken at times $(0, 1, \ldots, N)(1 + \tau(t))T_S$ instead of $(0, 1, \ldots, N)T_S$ where $\tau(t)$ is the process describing the clock jitter. The effect of sampling clock jitter is worse for higher analog frequencies. Suppose we are sampling a real sinusoidal signal $s(t) = A \sin(2\pi f t)$ with some jitter (Figure 3.30). Sampling frequency jitter results in an amplitude error. This error is the worst at the points where the slope of $s(t)$ is the highest, namely where $s(t)$ crosses the time axis. At these points, $ds/dt = 2\pi A f$. If the RMS jitter

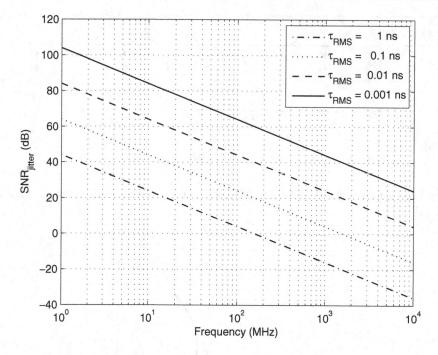

Figure 3.31 Impact of timing jitter on the SNR, for various τ_{RMS} of practical interest.

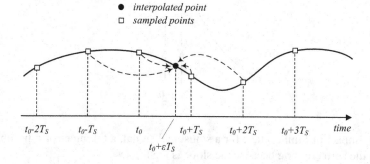

Figure 3.32 Samples around t_0 are used to compute the interpolated value.

is τ_{RMS}, the RMS amplitude error will be $2\pi A f \tau_{RMS}$, with the result that the SNR in the presence of sampling clock jitter is limited to:

$$SNR_{jitter} = 20\log_{10}\left(\frac{A}{2\pi A f \tau_{RMS}}\right) = -20\log_{10}(2\pi f \tau_{RMS}) \quad (3.53)$$

Very importantly, the frequency f in Equation (3.53) is the carrier frequency, not necessarily the symbol frequency. In the case of the digital quadrature generation described in

Section 3.1.4, f_{IF} must be used in Equation (3.53) to calculate the SNR. The impact of jitter on the SNR, as expressed by Equation (3.53), is plotted in Figure 3.31. As the frequency of the sampled carrier increases, the SNR drops rapidly. For example, to sample a signal centered at an IF frequency of 80 MHz with an SNR limited by jitter of 30 dB requires a maximum τ_{RMS} of about 50 ns. If the same signal is centered at 800 MHz, the maximum τ_{RMS} becomes 5 ps, which requires a more careful design and a better oscillator clock.

In order to model SCO and clock jitter in a sampled system, it is necessary to interpolate between the input samples. If the incoming sequence $x(t)$ is a sampled sequence at rate $1/T_S$ around time t_0, the interpolated value at time $t_0 + \epsilon T_S$ is given by:

$$x(t_0 + \epsilon T_S) = \sum_{k=n_1}^{n_2} x(t_0 + kT_S) \cdot p(\epsilon T_S - kT_S) \qquad (3.54)$$

where $p(t)$ is an interpolation polynomial, which determines the smoothness of the interpolation. As illustrated in Figure 3.32, few samples around t_0 are used to compute the interpolated value. For example, in the linear interpolator, $\{n_1, n_2\} = \{0, 1\}$ and $p(t)$ in Equation (3.54) takes values $\{1 - \epsilon, \epsilon\}$. The reader is referred to Erup et al. (1993) for a detailed treatment of interpolators. An alternate, accurate and simple manner to perform interpolation is to oversample (and filter) the signal and perform a simple linear interpolation over the oversampled sequence.

References

Erup, L., Gardner, F. and Harris, R. (1993) Interpolation in digital modems – part II: Implementation and performance. *IEEE Transactions on Communications* **41**(6), 998–1008.

Gard, K., Gutierrez, H. and Steer, M. (1999) Characterization of spectral regrowth in microwave amplifiers. *IEEE Transactions on Microwave Theory and Techniques* **47**, 1059–1069.

Manassevitch, V. (1962) *Frequency Synthesis Theory and Design*. John Wiley & Sons.

Meyr, H., Moeneclaey, M. and Fechtel, S. (1998) *Digital Communication Receivers Synchronization, Channel Estimation and Signal Processing*. John Wiley & Sons.

Nsenga, J., Thillo, W.V., Bourdoux, A., Ramon, V., Horlin, F. and Lauwereins, R. (2008) Spectral regrowth analysis of band-limited offset-QPSK. *IEEE Proceedings of International Conference on Acoustics, Speech and Signal Processing*.

Rapp, C. (1991) Effects of HPA-nonlinearity on an 4-DPSK/OFDM-signal for a digital sound broadcasting system. *Proceedings 2nd European Conf. On Satellite Communications*, pp. 176–184.

Razavi, B. (1998) *RF Microelectronics*. Prentice-Hall.

Saleh, A. (1981) Frequency-independent and frequency-dependent nonlinear models of TWT amplifiers. *IEEE Transactions on Communications* **29**, 1715–1720.

Vaughan, R., Scott, N. and White, D.R. (1991) The theory of bandpass sampling. *IEEE Transactions on Signal Processing* **39**(9), 1973–1984.

4

Impact of the Non-Ideal Front-Ends on the System Performance

Because of the limited frequency bandwidth, on the one hand, and the limited battery autonomy of terminal stations, on the other hand, spectral and power efficiencies of future communication systems should be as high as possible. New air interfaces, often derived from OFDM, are considered to meet the new requirements. Though OFDM is robust against the interference caused by the multi-path propagation, it is sensitive to the system non-idealities that destroy the orthogonality between the sub-carriers.

The CFO and the SCO, caused by the local oscillators at the transmitter and receiver, and the IQ imbalance, caused by the use of analog quadrature generation, are especially destructive. Since CFO, SCO and IQ imbalance are constant over each block of symbols (compared, for example, to additive noise or phase noise), it is important to study their effect based on a well-established model to assess if they can be compensated. Sections 4.1 and 4.2 are devoted to the study of the impact of CFO, SCO and IQ imbalance on the OFDM and SC-FDE air interfaces. The CFO and SCO can be equivalently defined at the transmitter or at the receiver. We limit the analysis to the IQ imbalance generated at the receiver (both the transmit and the receive IQ imbalance have a similar effect, except that the transmit IQ imbalance is generated before the channel convolution, making the performance degradation smaller in that case – see the analysis of the multi-user system). The performance degradation caused by phase noise, clipping and quantization noise and amplifier nonlinearity on the OFDM and SC-FDE systems is successively evaluated in Sections 4.4, 4.5 and 4.6. Figure 4.1 provides a block diagram of the single dimensional system (single antenna, single user) in the presence of the different front-end non-idealities.

The analysis is extended afterwards to multi-dimensional systems (multiple antennas, multiple users) that may suffer from additional interference between the different dimensions caused by the non-ideal front-end. Section 4.7 is devoted to multi-dimensional MIMO systems. Figure 4.40 provides a block diagram of the MIMO system considered in the third part of this chapter. Finally, multi-user systems are studied in Section 4.8. Figures 4.48 and 4.51 provide a block diagram of the downlink and uplink multi-user systems.

Digital Compensation for Analog Front-Ends François Horlin and André Bourdoux
© 2008 John Wiley & Sons, Ltd

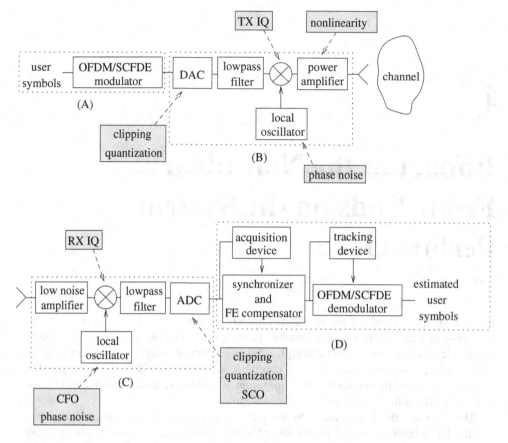

Figure 4.1 OFDM/SC-FDE system block diagram: (A) transmit digital transceiver; (B) transmit analog front-end; (C) receive analog front-end; (D) receive digital transceiver.

4.1 OFDM system in the presence of carrier frequency offset, sample clock offset and IQ imbalance

The impact of the CFO and SCO has been extensively addressed in the literature, even for multi-dimensional systems such as MC-CDMA. Pollet *et al.* (1995) first demonstrated that OFDM systems are very sensitive to CFO as they can only tolerate frequency offsets which are in a small fraction of the spacing between the sub-carriers without a large degradation in system performance. While Tomba and Krzymien (1999) evaluate analytically the impact of CFO on the performance of a single-user MC-CDMA system, Jang and Lee (1999) and Kim *et al.* (1999) extend the study of Tomba and Krzymien (1999) to a multi-user MC-CDMA system. Interestingly, Steendam and Moeneclaey (1999a) generalize the conclusion of Pollet *et al.* by showing that the performance of MC-CDMA systems depends strongly on the number of sub-carriers (or equivalently on the inter-carrier spacing). Steendam and Moeneclaey (1999a) study also the effects of carrier phase jitter and timing jitter on the

IMPACT OF THE NON-IDEAL FRONT-ENDS

system. Because the CFO and SCO cause a phase rotation linearly increasing with the symbol index, it is shown that the errors can be tracked based on the observation of the rotation on pilot symbols. Steendam and Moeneclaey (1999b, 2002) extend their conclusions to generalized MC-CDMA systems.

Even if the impact of IQ imbalance on OFDM systems has already been investigated by Tubbax *et al.* (2003), the sensitivity of multi-dimensional systems to this effect is not well addressed. The present section studies the joint effect of CFO, SCO and IQ imbalance on a OFDM-based communication system (Horlin *et al.* 2006).

4.1.1 Model of the non-idealities in the frequency domain

System model

The purpose of this section is to build an equivalent model of the CFO, SCO and IQ imbalance in the frequency domain. The system under consideration is represented in Figure 4.2. Note that we focus on one symbol block, and that the block index has been discarded for the sake of clarity.

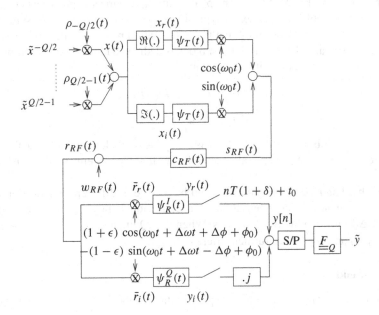

Figure 4.2 Model of the non-idealities (CFO: $\Delta\omega$; SCO: δ; IQ imbalance: ϵ, $\Delta\phi$, $\psi_R^I(t)$ and $\psi_R^Q(t)$). Reproduced by permission of ©2006 IEEE [Horlin *et al*].

Each symbol $\tilde{x}^q$ in the symbol vector $\tilde{x}$, defined as $\tilde{x} := [\tilde{x}^{-Q/2} \ldots \tilde{x}^{Q/2-1}]^T$, is transmitted on the sub-carrier frequency q by applying an IFFT on $\tilde{x}$. The occupied frequency spectrum has been centered around the carrier frequency. The symbol block duration is equal to QT, where Q is the number of sub-carriers and T is the duration of one symbol. Assuming that the vector of symbols is sent periodically (which is performed in practice by the addition of a cyclic prefix at the transmitter and by the removal of the corresponding samples at the receiver), the IFFT is equivalent to multiplying $\tilde{x}^q$ by $\rho_q(t) = (1/\sqrt{Q})\, e^{j2\pi qt/QT}$ and summing the contributions on the different sub-carriers. The baseband signal is up-converted to the carrier frequency ω_0, after lowpass filtering by $\psi_T(t)$ in order to remove the out-of-band components. The resulting signal $s_{RF}(t)$ is transmitted through a frequency selective channel $c_{RF}(t)$. Additive white Gaussian noise $w_{RF}(t)$, of one-sided power spectral density equal to N_0, is added in the first amplifier stages of the received front-end. The RF received signal $r_{RF}(t)$ is finally down-converted to the baseband domain for complex operation, lowpass filtered in order to avoid aliasing and sampled before going back to the frequency domain by the use of a FFT, corresponding to the multiplication of the received vector of samples with a matrix $\underline{\underline{F}}_Q$. The received vector is composed by the signals received on the different sub-carrier frequencies, as defined in $\tilde{y} := [\tilde{y}^{-Q/2} \ldots \tilde{y}^{Q/2-1}]^T$.

Due to the fact that the local oscillator at the receive terminal is different from the one at the transmit base-station, the down-conversion to the baseband domain is operated with a phase shift ϕ_0 and with a frequency shift $\Delta\omega$ (CFO), and the received signal is sampled with an initial phase t_0 different from 0 and with a period T' slightly different from the one at the transmitter T, $T' = T(1 + \delta)$ (SCO). The phase shift ϕ_0 and the sampling instant t_0 are both functions of the symbol block index. A typical burst composed of multiple symbol blocks is illustrated in Figure 4.3. Assuming that the initialization point is in the middle of the first block (which is typically the case if the first block is dedicated to the channel estimation) and focusing on the symbol block n, it is clear that:

$$\phi_0 = (n-1)(Q + L_{cp})T\Delta\omega \quad (4.1)$$

$$t_0 = (n-1)(Q + L_{cp})T\delta \quad (4.2)$$

in which Q is the number of carriers and L_{cp} is the cyclic prefix length. Note that the sampling phase drift can overlay one sample after a while and that the system should compensate for this effect.

On the other hand, IQ imbalance is caused by the use of different elements on the I and Q branches. Frequency-independent IQ imbalance is modeled by a difference in amplitude ϵ and phase $\Delta\phi$ between the two branches. When the IQ imbalance is frequency-dependent, the filters $\psi_R^I(t)$ and $\psi_R^Q(t)$ are further different.

Receive front-end

If $r(t) := r_r(t) + jr_i(t)$ is the baseband representation of the RF received signal $r_{RF}(t)$, we have that:

$$r_{RF}(t) = r_r(t)\cos(\omega_0 t) - r_i(t)\sin(\omega_0 t) \quad (4.3)$$

Taking the CFO and IQ imbalance into account, the signal on the I branch after multiplication by the cosine ($\times 2$) is:

$$\bar{r}_r(t) = r_r(t)(1+\epsilon)\cos(\Delta\phi + \Delta\omega t + \phi_0) + r_i(t)(1+\epsilon)\sin(\Delta\phi + \Delta\omega t + \phi_0) \quad (4.4)$$

and the signal on the Q branch after multiplication by the sine (sign inversion, $\times 2$) is:

$$\bar{r}_i(t) = r_r(t)(1-\epsilon)\sin(\Delta\phi - \Delta\omega t - \phi_0) + r_i(t)(1-\epsilon)\cos(\Delta\phi + \Delta\omega t + \phi_0) \quad (4.5)$$

IMPACT OF THE NON-IDEAL FRONT-ENDS

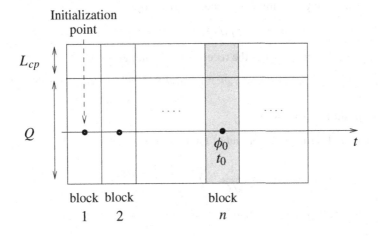

Figure 4.3 Phase shift ϕ_0 and sampling phase t_0.

assuming that the high frequency components are later filtered out by the filter $\psi_R(t)$ ($\psi_R(t) := \psi_R^I(t) = \psi_R^Q(t)$ in the case of frequency independent IQ imbalance). By defining $\bar{r}(t) := \bar{r}_r(t) + j\bar{r}_i(t)$, we obtain:

$$\bar{r}(t) = \alpha\, e^{-j(\Delta\omega t + \phi_0)} r(t) + \beta\, e^{j(\Delta\omega t + \phi_0)} r^*(t) \qquad (4.6)$$

in which:

$$\alpha := \cos(\Delta\phi) - j\epsilon \sin(\Delta\phi) \qquad (4.7)$$

$$\beta := \epsilon \cos(\Delta\phi) + j \sin(\Delta\phi) \qquad (4.8)$$

The parameters α and β express the impact of the IQ imbalance on the received signal as in Section 3.3.5 ($\alpha = 1$ and $\beta = 0$ when the two paths are exactly in quadrature). The CFO leads to a frequency shift of the two components of the baseband received signal, as shown in Equation (4.6). At the output of the lowpass filter $\psi_R(t)$, we obtain:

$$y(t) = \bar{r}(t) \star \psi_R(t) \qquad (4.9)$$

$$= \alpha\, e^{-j(\Delta\omega t + \phi_0)} (r(t) \star \psi_R(t)\, e^{j\Delta\omega t}) + \beta\, e^{j(\Delta\omega t + \phi_0)} (r(t) \star \psi_R(t)\, e^{j\Delta\omega t})^* \qquad (4.10)$$

in which $\star$ represents the convolution operator.

Time domain communication

If $x(t)$ denotes the time domain baseband signal at the input of the transmit front-end, and if $c(t)$ and $w(t)$ denote the baseband representations of the RF channel $c_{RF}(t)$ and noise $w_{RF}(t)$, it is clear that:

$$r(t) = x(t) \star (\psi_T(t) \star c(t)) + w(t) \qquad (4.11)$$

and, based on Equation (4.10), we finally obtain:

$$y(t) = \alpha\, e^{-j(\Delta\omega t + \phi_0)} (x(t) \star h(t)) + \beta\, e^{j(\Delta\omega t + \phi_0)} (x(t) \star h(t))^* + z(t) \qquad (4.12)$$

in which $h(t)$ is the composite impulse response, given by:

$$h(t) := \psi_T(t) \star c(t) \star \psi_R(t) \, e^{j\Delta\omega t} \qquad (4.13)$$

and $z(t)$ is the noise at the output of the receiver front-end, given by:

$$z(t) := \alpha \, e^{-j(\Delta\omega t + \phi_0)} (w(t) \star \psi_R(t) \, e^{j\Delta\omega t}) + \beta \, e^{j(\Delta\omega t + \phi_0)} (w(t) \star \psi_R(t) \, e^{j\Delta\omega t})^* \qquad (4.14)$$

Frequency domain transmission

Since the transmitted signal at the output of the IFFT is given by:

$$x(t) = \frac{1}{\sqrt{Q}} \sum_{q=-Q/2}^{Q/2-1} \tilde{x}^q \, e^{j2\pi qt/QT} \qquad (4.15)$$

based on Equation (4.12), we have that:

$$y(t) = \alpha \frac{1}{\sqrt{Q}} \sum_{q=-Q/2}^{Q/2-1} \tilde{x}^q \tilde{h}^q \, e^{(j2\pi qt/QT)} \, e^{-j(\Delta\omega t + \phi_0)}$$

$$+ \beta \frac{1}{\sqrt{Q}} \sum_{q=-Q/2}^{Q/2-1} (\tilde{x}^q)^* (\tilde{h}^q)^* \, e^{-j(2\pi qt/QT)} \, e^{j(\Delta\omega t + \phi_0)} + z(t) \qquad (4.16)$$

in which $\tilde{h}^q$ is the composite impulse response $h(t)$ at the carrier frequency q. After sampling, we obtain:

$$y[n] := y(t = nT(1+\delta) + t_0) \qquad (4.17)$$

$$= \alpha \frac{1}{\sqrt{Q}} \sum_{q=-Q/2}^{Q/2-1} \tilde{x}^q \tilde{h}^q \, e^{-j(\phi_0 + \Delta\omega t_0)} \cdot e^{j2\pi t_0/QTq} \, e^{j(2\pi q/Q - \Delta\omega T)(1+\delta)n}$$

$$+ \beta \frac{1}{\sqrt{Q}} \sum_{q=-Q/2}^{Q/2-1} (\tilde{x}^q)^* (\tilde{h}^q)^* \, e^{j(\phi_0 + \Delta\omega t_0)} \cdot e^{-j(2\pi t_0/QTq)} \, e^{-j(2\pi q/Q - \Delta\omega T)(1+\delta)n} + z[n]$$

$$(4.18)$$

where $z[n] := z(t = nT(1+\delta) + t_0)$. Assuming that the filter $\psi_R(t)$ is a perfect lowpass filter of bandwidth $1/T'$, the noise samples are independent and of variance equal to $2N_0/T'(1+\epsilon^2) \simeq 2N_0/T'$.

Frequency domain reception

At the output of the FFT, the signal on each carrier p is:

$$\tilde{y}^p = \frac{1}{\sqrt{Q}} \sum_{n=-Q/2}^{Q/2-1} y[n] \, e^{-j(2\pi np/Q)} \qquad (4.19)$$

$$= \alpha \sum_{q=-Q/2}^{Q/2-1} \tilde{x}^q \tilde{h}^q \gamma_0^q \gamma^{p,q} + \beta \sum_{q=-Q/2}^{Q/2-1} (\tilde{x}^q)^* (\tilde{h}^q)^* (\gamma_0^q)^* (\gamma^{-p,q})^* + \tilde{z}^p \qquad (4.20)$$

IMPACT OF THE NON-IDEAL FRONT-ENDS

in which the factors γ_0^q and $\gamma^{p,q}$ are defined such that:

$$\angle(\gamma_0^q) := \left(-\phi_0 + \frac{\Delta\omega T}{2}\right) + \left(-\Delta\omega t_0 - \frac{\Delta\omega T\delta}{2}\right) + \left(\frac{2\pi t_0}{QT} - \frac{\pi\delta}{Q}\right)q \quad (4.21)$$

$$|\gamma_0^q| := \frac{1}{Q} \frac{\sin(\frac{\Delta\omega QT}{2} + \frac{\Delta\omega QT\delta}{2} - \pi q\delta)}{\sin(\frac{\Delta\omega T}{2} + \frac{\Delta\omega T\delta}{2} - \frac{\pi q\delta}{Q})} \quad (4.22)$$

$$\gamma^{p,q} := \frac{(-1)^{(q-p)} \cdot e^{-j(\pi(q-p)/Q)}}{\cos[\frac{\pi(q-p)}{Q}] - \frac{1}{\tan(\frac{\Delta\omega T}{2} + \frac{\Delta\omega T\delta}{2} - \frac{\pi q\delta}{Q})} \sin(\frac{\pi(q-p)}{Q})} \quad (4.23)$$

and $\tilde{z}^p := (1/\sqrt{Q})\sum_{n=-Q/2}^{Q/2-1} z[n] e^{-j2\pi np/Q}$ is the noise in the frequency domain. Equation (4.20) has been obtained by noting that $\sum_{n=0}^{Q-1}(x)^n = (1-x^Q)/(1-x)$.

The expression (4.20) can be summarized with the following matrix model:

$$\underline{\tilde{y}} = \alpha \underline{\underline{\gamma}} \cdot \underline{\underline{\Lambda}}_0 \cdot \underline{\underline{\Lambda}}_{\tilde{h}} \cdot \underline{\tilde{x}} + \beta \underline{\underline{\tilde{\gamma}}}^* \cdot \underline{\underline{\Lambda}}_0^* \cdot \underline{\underline{\Lambda}}_{\tilde{h}}^* \cdot \underline{\tilde{x}}^* + \underline{\tilde{z}} \quad (4.24)$$

where the matrix $\underline{\underline{\gamma}}$ is equal to $\underline{\underline{\gamma}} := [\gamma^{p,q}]_{p,q=-Q/2\cdots Q/2-1}$ and the matrix $\underline{\underline{\tilde{\gamma}}}$ is a mirrored version of $\underline{\underline{\gamma}}$ (the rows corresponding to $p = -Q/2+1, \ldots, Q/2-1$ are flipped). The matrices $\underline{\underline{\Lambda}}_0$ and $\underline{\underline{\Lambda}}_{\tilde{h}}$ are diagonal. Matrix $\underline{\underline{\Lambda}}_0$ contains the elements γ_0^q on its diagonal. Matrix $\underline{\underline{\Lambda}}_{\tilde{h}}$ contains the channel coefficients $\tilde{h}^q$ on its diagonal. The noise vector $\underline{\tilde{z}}$ is equal to $\underline{\tilde{z}} := [\tilde{z}^{-Q/2} \ldots \tilde{z}^{Q/2-1}]^T$.

CFO and SCO cause first the phase shift (4.21) and the amplitude distortion (4.22) on each carrier independently, and second interference between the carriers (4.23). IQ imbalance produces supplementary interference between the carriers (term in β in Equation (4.20)). In the following sections, Equation (4.24) is particularized for each effect independently. The combined impact of the different non-idealities is studied afterwards.

4.1.2 Effect of carrier frequency offset

When the system suffers only from CFO (no SCO and no IQ imbalance), the expression (4.24) becomes:

$$\underline{\tilde{y}} = \underline{\underline{\gamma}} \cdot \underline{\underline{\Lambda}}_0 \cdot \underline{\underline{\Lambda}}_{\tilde{h}} \cdot \underline{\tilde{x}} + \underline{\tilde{z}} \quad (4.25)$$

where the constituting terms of the matrices $\underline{\underline{\Lambda}}_0$ and $\underline{\underline{\gamma}}$ are given by:

$$\angle(\gamma_0^q) = -\phi_0 + \frac{\Delta\omega T}{2} \quad (4.26)$$

$$|\gamma_0^q| = \frac{1}{Q} \frac{\sin(\frac{\Delta\omega QT}{2})}{\sin(\frac{\Delta\omega T}{2})} \quad (4.27)$$

$$\gamma^{p,q} = \frac{(-1)^{(q-p)} \cdot e^{-j\pi(q-p)/Q}}{\cos(\frac{\pi(q-p)}{Q}) - (\tan(\frac{\Delta\omega T}{2}))^{-1} \sin(\frac{\pi(q-p)}{Q})} \quad (4.28)$$

The effects of CFO are therefore:

- the phase rotation (4.26) common to all carriers;
- the signal amplitude distortion (4.27) common to all carriers;
- interference of each carrier on the symbols located on its neighboring carriers (4.28).

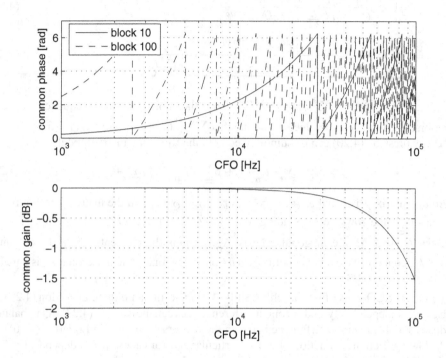

Figure 4.4 Common phase and gain as a function of the CFO.

The common rotation is composed of the dominating reference phase ϕ_0 in the middle of the block, plus a smaller correction term corresponding to the phase drift during half a sample. The upper part of Figure 4.4 illustrates the common phase as a function of the CFO for two different blocks (blocks 10 and 100 in the burst). Even for small values of the CFO, the common phase is rapidly significant and has to be compensated. To this end, pilot symbols are often inserted in the block based on which the common phase rotation can be estimated. In the next simulation results, we assume that the common phase rotation has been perfectly compensated. The common gain corresponds to a loss in signal power in favor of the ICI power. The lower part of Figure 4.4 shows that the loss is generally small (less than 0.5 dB for a CFO smaller than 50 kHz).

The performance of the OFDM system affected by CFO is illustrated in Figure 4.5 for the AWGN channel. The left part of the figure shows the ratio of the signal power (the diagonal elements of the matrix $\overline{\overline{\gamma}}$) to the interference power (the non-diagonal elements of the matrix $\overline{\overline{\gamma}}$). The right part of the figure shows the resulting BER as a function of the SNR, assuming that the common phase drift has been successfully compensated. Because the inter-carrier interference (ICI) can be well approximated by an additive Gaussian process, it is possible to link the signal-to-interference ratio (SIR) to the asymptotic BER performance. As an example, the SIR is limited to 15 dB for a CFO equal to 30 kHz. Therefore, the BER will floor at approximately 2×10^{-3} which corresponds to the value of the BER for an SNR equal to 15 dB when there is no CFO in the system.

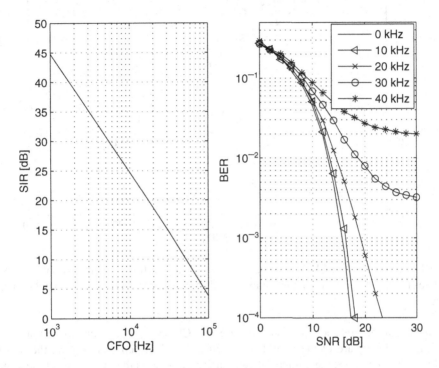

Figure 4.5 Impact of the CFO (only ICI) on the system performance in case of the AWGN channel.

Finally, the impact of the CFO on the system performance is studied in Figure 4.6 for frequency selective channels. We compare the BER performance achieved when the channels are weakly frequency selective (the solid curves correspond to the 10% channels having the lowest delay spread) to the performance achieved when the channels are highly frequency selective (the dashed curves correspond to the 10% channels having the highest delay spread). It can be generally stated that the impact of the CFO is similar in the two cases even if the performance is worse for highly selective channels. It can be shown that the SIR averaged over the sub-carriers for any frequency selective channel is near to the SIR computed for the AWGN channel. It is, however, difficult to deduce the BER asymptotic performance from the average SIR, since the performance is mostly limited by the worst carriers. To conclude, the residual CFO after compensation should be smaller than 10 kHz to obtain a sufficiently small performance degradation.

As a final remark, we would like to draw the attention of the reader to the influence of the number of sub-carriers, or equivalently of the inter-carrier spacing, on the system

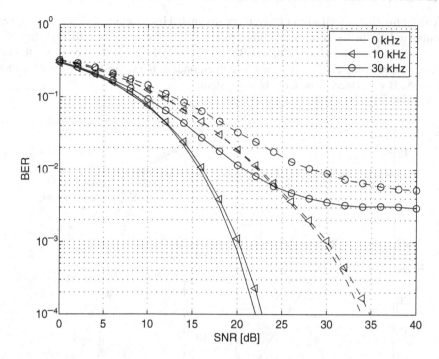

Figure 4.6 Impact of the CFO (only ICI) on the system performance in case of frequency selective channels (solid curves: low delay spread channels; dashed curves: high delay spread channels).

performance in the presence of CFO. Because the ICI power (one minus the signal power loss equal to the square of the modulus of Equation (4.28)) depends only on the product $\Delta\omega QT$ in the first order, the number of sub-carriers influences the sensitivity of the system to the CFO proportionally. Considering, for example, the 3GPP LTE system in which the inter-carrier spacing is approximately 10 times smaller than the WLAN system considered in this section, the required CFO estimation/compensation accuracy should be improved by a factor 10.

4.1.3 Effect of sample clock offset

When the system suffers only from SCO (no CFO and no IQ imbalance), the expression (4.24) becomes:

$$\underline{\tilde{y}} = \underline{\underline{\gamma}} \cdot \underline{\underline{\Lambda}}_0 \cdot \underline{\underline{\Lambda}}_{\tilde{h}} \cdot \underline{\tilde{x}} + \underline{\tilde{z}} \qquad (4.29)$$

where the constituting terms of the matrices $\underline{\underline{\Lambda}}_0$ and $\underline{\underline{\gamma}}$ are given by:

$$\angle(\gamma_0^q) := \left(\frac{2\pi t_0}{QT} - \frac{\pi \delta}{Q} \right) q \qquad (4.30)$$

$$|\gamma_0^q| := \frac{1}{Q} \frac{\sin(\pi q \delta)}{\sin(\frac{\pi q \delta}{Q})} \qquad (4.31)$$

IMPACT OF THE NON-IDEAL FRONT-ENDS

$$\gamma^{p,q} := \frac{(-1)^{(q-p)} \dot{e}^{-j\pi(q-p)/Q}}{\cos(\frac{\pi(q-p)}{Q}) + (\tan(\frac{\pi q \delta}{Q}))^{-1} \sin(\frac{\pi(q-p)}{Q})} \qquad (4.32)$$

The effects of SCO are therefore:

- the phase rotation (4.30) proportional to the carrier index;
- the signal amplitude distortion (4.31) common to all carriers;
- interference of each carrier on the symbols located on its neighboring carriers (4.32).

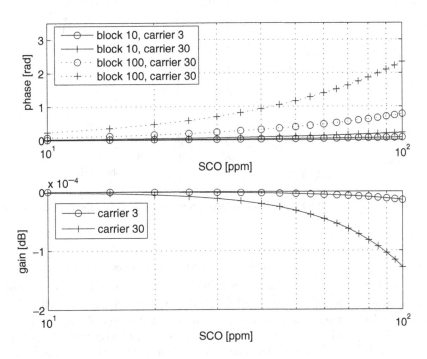

Figure 4.7 Phase and gain as a function of the SCO.

The rotation is due to the dominating reference sampling time instant t_0 in the middle of the block, plus a small correction term corresponding to the sampling time drift during half a sample. The upper part of Figure 4.7 illustrates the phase as a function of the SCO for two different sub-carriers (sub-carriers 3 and 30 in the block) and two different blocks (blocks 10 and 100 in the burst). Even if the phase drift generated by the SCO on the constellation is much smaller than the phase drift generated by the CFO, it is also particularly destructive and should be compensated before the symbol detection can be performed. In the next results, we assume that the phase drift has been perfectly compensated. The signal power distortion illustrated in the lower part of Figure 4.4 corresponds to a loss of signal power in favor of the ICI power and is, on the whole, negligible (very close to 0 dB).

The performance of the system affected by SCO is illustrated in Figure 4.8 for the AWGN channel. The left part of the figure shows the ratio of the signal power (the diagonal

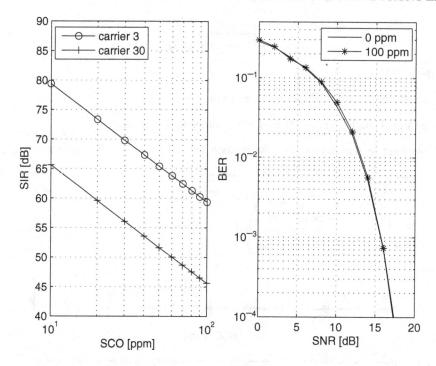

Figure 4.8 Impact of the SCO (only ICI) on the system performance in the case of the AWGN channel.

elements of matrix $\underline{\underline{\gamma}}$) to the interference power (the non-diagonal elements of the matrix $\underline{\underline{\gamma}}$). The same sub-carrier indexes have been considered (sub-carriers 3 and 30 in the block). The right part of the figure shows the resulting BER as a function of the SNR, assuming that the constellation phase drift has been successfully compensated. For reasonable values of the SCO (100 ppm clock drift corresponds already to a very bad crystal), the SIR is high enough and the BER performance degradation is negligible. This conclusion is generalized in Figure 4.9 for frequency-selective channels.

As a final remark, we like to mention that the number of sub-carriers, or equivalently the inter-sub-carrier spacing, influences the system performance in the presence of SCO in a similar fashion as in the presence of CFO.

4.1.4 Effect of IQ imbalance

When the system suffers only from IQ imbalance (no CFO and no SCO), the expression (4.24) becomes:

$$\underline{\tilde{y}} = \alpha \underline{\underline{\Lambda}}_{\tilde{h}} \cdot \underline{\tilde{x}} + \beta \underline{\underline{\tilde{I}}}_Q \cdot \underline{\underline{\Lambda}}_{\tilde{h}}^* \cdot \underline{\tilde{x}}^* + \underline{\tilde{z}} \qquad (4.33)$$

where the product with the mirrored identity matrix $\underline{\underline{\tilde{I}}}_Q = \underline{\underline{F}}_Q \cdot \underline{\underline{F}}_Q$ reduces to a mirror operation. Therefore, IQ imbalance adds mainly an image of the transmitted vector (complex conjugated and mirrored) that interferes with the desired signal.

IMPACT OF THE NON-IDEAL FRONT-ENDS 83

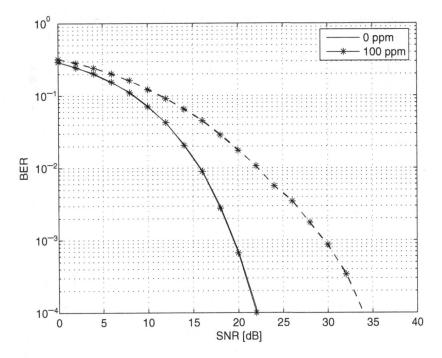

Figure 4.9 Impact of the SCO (only ICI) on the system performance in case of frequency selective channels (solid curves: low-delay spread channels; dashed curves: high-delay spread channels).

Figure 4.10 illustrates the parameters α and β as a function of the phase mismatch for different values of the amplitude mismatch. The signal gain α is nearly independent of the IQ imbalance. The interference gain β, on the other hand, depends heavily on the IQ imbalance. The phase mismatch (evaluated as a percentage of 2π) and the amplitude mismatch (evaluated as a percentage of 1) have approximately the same effect.

The performance of the system affected by IQ imbalance in the case of the AWGN channel is given in Figure 4.11. The left part of the figure illustrates the SIR as a function of the phase mismatch (assuming no amplitude mismatch). The right part of the figure illustrates the BER performance for different values of the phase mismatch. The IQ imbalance cannot be assimilated to a Gaussian noise process because each carrier suffers only from the interference coming from another carrier (compared to CFO and SCO generating interference from many carriers). There is therefore no direct correspondence between the SIR and the asymptotic BER behavior.

Finally, Figure 4.12 evaluates the performance of the system in the case of frequency selective channels. Weak-frequency selective channels (10% lowest delay spread channels) are compared to high-frequency selective channels (10% highest delay spread channels). The impact of IQ imbalance increases with the frequency selectivity of the channels. Furthermore, the impact of IQ imbalance in the case of high-frequency selective channels is dramatic: the BER curve floors already at 10^{-2} for a 1% phase mismatch. This can be understood

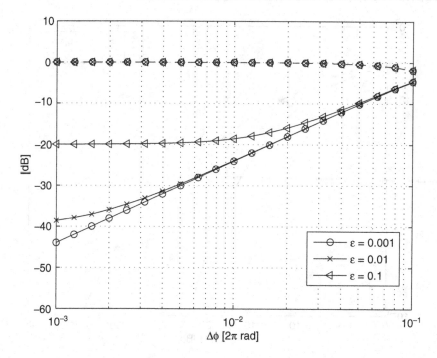

Figure 4.10 Impact of the IQ imbalance on the parameters α (dashed curves) and β (solid curves).

by the fact that a high-gain carrier can interfere on a low-gain carrier located at the other side of the spectrum destroying completely the symbol detection on that carrier. To conclude, the residual phase and amplitude mismatch after compensation should be much smaller than 10^{-2} to make sure that the system can work.

4.1.5 Combination of effects

When the CFO/SCO/IQ imbalance non-idealities are studied jointly, common terms that have been previously neglected are taken into account. For example, we observe:

- a supplementary common phase rotation (second term in Equation (4.21)) and a modification of the ICI power (second term in the tangent of Equation (4.23)) due to the joint presence of CFO and SCO;

- a phase rotation and a spreading over multiple neighboring carriers of the interference generated by the IQ imbalance due to the CFO and SCO (matrices $\underline{\underline{\Lambda}}_0$ and $\underline{\underline{\tilde{\gamma}}}$ in the term in β of Equation (4.24)).

The resulting model (4.24) is, however, hardly tractable and should be simplified. It is possible to make the following approximations:

- the common rotation is caused mainly by the reference phase and sampling instant in the block (and not by the supplementary small correction terms);

IMPACT OF THE NON-IDEAL FRONT-ENDS

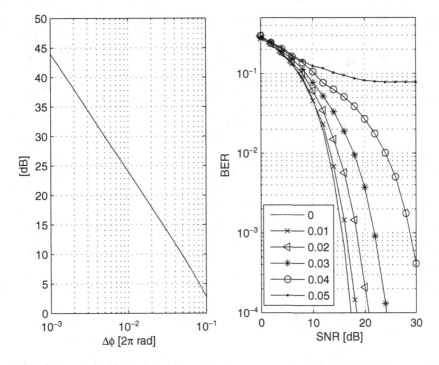

Figure 4.11 Impact of the phase mismatch on the system performance in the case of the AWGN channel.

- the signal gain is approximately equal to 1 (see Figures 4.4, 4.7 and 4.10);
- the ICI is caused mainly by the CFO (interference on the neighboring carriers) and by the IQ imbalance (interference on the mirror carrier) independently.

Therefore, the frequency domain model (4.24) can be interestingly approximated by:

$$\underline{\tilde{y}} = \underline{\underline{\gamma}} \cdot \underline{\underline{\Lambda}}_0 \cdot \underline{\underline{\Lambda}}_{\tilde{h}} \cdot \underline{\tilde{x}} + \beta \underline{\tilde{I}}_Q \cdot \underline{\underline{\Lambda}}_0^* \cdot \underline{\underline{\Lambda}}_{\tilde{h}}^* \cdot \underline{\tilde{x}}^* + \underline{\tilde{z}} \qquad (4.34)$$

where $\underline{\underline{\Lambda}}_0$ is a diagonal matrix composed of elements of modulus 1 and of phase given by:

$$\angle(\gamma_0^q) := -\phi_0 - \Delta\omega t_0 + \frac{2\pi t_0}{QT} q \qquad (4.35)$$

and $\underline{\underline{\gamma}}$ is the ICI matrix composed of the elements:

$$\gamma^{p,q} := \frac{(-1)^{(q-p)} \dot{e}^{-j(\pi(q-p)/Q)}}{\cos(\frac{\pi(q-p)}{Q}) - (\frac{\Delta\omega T}{2})^{-1} \sin(\frac{\pi(q-p)}{Q})} \qquad (4.36)$$

In the last approximation, most of the terms common to multiple non-idealities have been neglected because they are of the second order (the remaining common terms are mainly

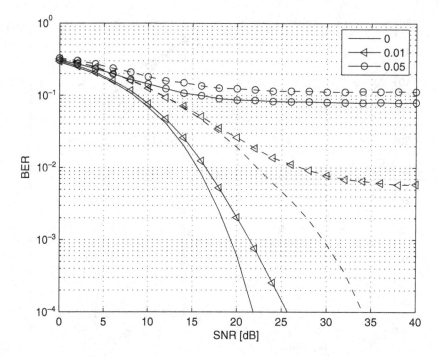

Figure 4.12 Impact of the phase mismatch on the system performance in the case of frequency selective channels (solid curves: low delay spread channels; dashed curves: high delay spread channels).

the phase rotation of the signal and of the interference). The approximation is verified in Figure 4.13 when the CFO, SCO and IQ imbalance are equal to their recommended specification (10 kHz CFO, 100 ppm SCO and 0.1% amplitude and phase mismatch). The approximated phase rotation, computed based on Equation (4.35), is very close to the actual phase rotation for all carriers. The approximated SIR, computed based on Equation (4.36), is very close to the actual SIR averaged over all sub-carriers (but the slight dependence on the sub-carrier index is neglected).

The system performance is finally evaluated in Figure 4.14 when the CFO, SCO and IQ imbalance are active at their specification. We study separately the performance in the case of the low-delay spread and high-delay spread channels. In both cases, the achieved degradation is acceptable. We will demonstrate in the next chapter that non-ideality estimation/compensation algorithms can be developed to meet those specifications.

4.1.6 Extension to the frequency-dependent IQ imbalance

For the sake of completeness, we extend the frequency domain model of the non-idealities to frequency-dependent IQ imbalance. Frequency-dependent IQ imbalance is generated by a difference between the amplifiers and filters on the I and Q branches of the analog front-end, resulting in a difference of impulse responses.

IMPACT OF THE NON-IDEAL FRONT-ENDS

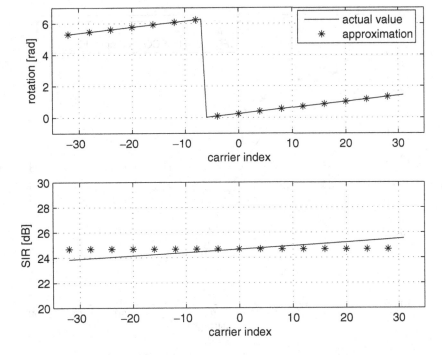

Figure 4.13 Verification of the model approximation; 10 kHz CFO, 100 ppm SCO, 0.1% amplitude and phase mismatch, AWGN channel, block 50.

In the mathematical description of the analog front-end, frequency-dependent IQ imbalance is modeled by a difference in the lowpass filters on the I and Q branches ($\psi_R^I(t) \neq \psi_R^Q(t)$). The signal $\bar{r}_r(t)$ in Equation (4.4) filtered out by the lowpass filter $\psi_R^I(t)$ is given by:

$$y_r(t) = \tfrac{1}{2}(1+\epsilon)(\cos(\Delta\phi) + j\sin(\Delta\phi))(\psi_R^I(t) \star e^{j(\Delta\omega t + \phi_0)} r^*(t))$$
$$+ \tfrac{1}{2}(1+\epsilon)(\cos(\Delta\phi) - j\sin(\Delta\phi))(\psi_R^I(t) \star e^{-j(\Delta\omega t + \phi_0)} r(t)) \quad (4.37)$$

and the signal $\bar{r}_i(t)$ in Equation (4.5) filtered out by the lowpass filter $\psi_R^Q(t)$ is given by:

$$y_i(t) = j\tfrac{1}{2}(1-\epsilon)(\cos(\Delta\phi) + j\sin(\Delta\phi))(\psi_R^Q(t) \star e^{-j(\Delta\omega t + \phi_0)} r(t))$$
$$- j\tfrac{1}{2}(1-\epsilon)(\cos(\Delta\phi) - j\sin(\Delta\phi))(\psi_R^Q(t) \star e^{j(\Delta\omega t + \phi_0)} r^*(t)) \quad (4.38)$$

By defining $y(t) := y_r(t) + jy_i(t)$, we obtain:

$$y(t) = e^{-j(\Delta\omega t + \phi_0)}\{r(t) \star (\alpha\psi_R(t) e^{j\Delta\omega t} + \beta^*\epsilon_R(t) e^{j\Delta\omega t})\}$$
$$+ e^{j(\Delta\omega t + \phi_0)}\{r(t) \star (\beta^*\psi_R(t) e^{j\Delta\omega t} + \alpha\epsilon_R(t) e^{j\Delta\omega t})\}^* \quad (4.39)$$

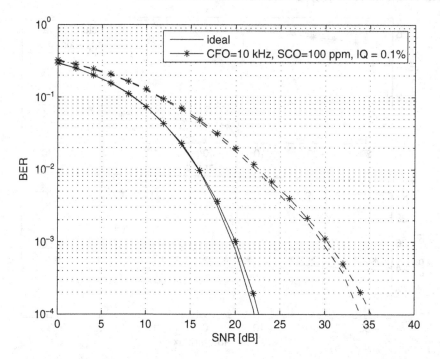

Figure 4.14 System performance in the presence of 10 kHz CFO, 100 ppm SCO and 0.1% amplitude and phase mismatch (solid curves: low-delay spread channels; dashed curves: high-delay spread channels).

in which α and β, defined in Equations (4.7) and (4.8), express the effect of the amplitude and phase mismatch generated in the local oscillator, and:

$$\psi_R(t) := (\psi_R^I(t) + \psi_R^Q(t))/2$$
$$\epsilon_R(t) := (\psi_R^I(t) - \psi_R^Q(t))/2 \qquad (4.40)$$

denote the average lowpass filter and the impulse response mismatch, respectively, generated in the amplifiers and filters.

Similarly to Equation (4.12), the signal $y(t)$ can be expressed as a function of the signal $x(t)$ at the input of the transmit front-end:

$$y(t) = e^{-j(\Delta\omega t + \phi_0)}(x(t) \star h_\alpha(t)) + e^{j(\Delta\omega t + \phi_0)}(x(t) \star h_\beta(t))^* + z(t) \qquad (4.41)$$

in which $h_\alpha(t)$ is the composite impulse response characterizing the direct path:

$$h_\alpha(t) := \psi_T(t) \star c(t) \star (\alpha\psi_R(t) e^{j\Delta\omega t} + \beta^* \epsilon_R(t) e^{j\Delta\omega t}) \qquad (4.42)$$

and $h_\beta(t)$ is the composite impulse response characterizing the mirror path:

$$h_\beta(t) := \psi_T(t) \star c(t) \star (\beta^* \psi_R(t) e^{j\Delta\omega t} + \alpha\epsilon_R(t) e^{j\Delta\omega t}) \qquad (4.43)$$

IMPACT OF THE NON-IDEAL FRONT-ENDS

The noise $z(t)$ at the output of the receiver front-end is equal to:

$$z(t) = e^{-j(\Delta\omega t + \phi_0)}(w(t) \star (\alpha \psi_R(t) e^{j\Delta\omega t} + \beta^* \epsilon_R(t) e^{j\Delta\omega t}))$$
$$+ e^{j(\Delta\omega t + \phi_0)}(w(t) \star (\beta^* \psi_R(t) e^{j\Delta\omega t} + \alpha \epsilon_R(t) e^{j\Delta\omega t}))^* \quad (4.44)$$

Based on Equation (4.41), the final frequency domain model (4.24) is easily adapted to encompass the frequency-dependent IQ imbalance:

$$\underline{\tilde{y}} = \underline{\gamma} \cdot \underline{\underline{\Lambda}}_0 \cdot \underline{\underline{\Lambda}}_{\tilde{h}_\alpha} \cdot \underline{\tilde{x}} + \underline{\tilde{\gamma}}^* \cdot \underline{\underline{\Lambda}}_0^* \cdot \underline{\underline{\Lambda}}_{\tilde{h}_\beta}^* \cdot \underline{\tilde{x}}^* + \underline{\tilde{z}} \quad (4.45)$$

in which $\underline{\underline{\Lambda}}_{\tilde{h}_\alpha}$ is a diagonal matrix containing the sampled composite impulse response $h_\alpha[n] := h_\alpha(t = nT)$ in the frequency domain on its diagonal, and $\underline{\underline{\Lambda}}_{\tilde{h}_\beta}$ is a diagonal matrix containing the sampled composite impulse response $h_\beta[n] := h_\beta(t = nT)$ in the frequency domain on its diagonal.

In most of the existing communication systems, the IQ imbalance can be assumed to be frequency-independent because the bandwidth is not too large (less than 20 MHz). It will, however, be shown in the next chapter that frequency-dependent IQ imbalance can also be estimated and compensated.

4.2 SC-FDE system in the presence of carrier frequency offset, sample clock offset and IQ imbalance

In the literature, the impact of CFO, SCO and IQ imbalance on the SC-FDE system is not well studied. The focus is rather on the design of pilot symbols in the time domain to enable the time and frequency synchronization of the SC-FDE system (see Yu et al. (2003) and Reinhardt and Weigel (2004)). The aim of this section is to assess the sensitivity of the SC-FDE system to the front-end non-ideality, and to compare it to the sensitivity of the OFDM system.

SC-FDE has been presented in Section 2.2 as a special case of a linearly pre-coded OFDM system (see the linearly pre-coded OFDM system in Figure 2.1 and the resulting SC-FDE system in Figure 2.2). At the transmitter, the FFT linear pre-coder compensates for the IFFT to make the signal envelope less variable over the time (lower PAPR):

$$\underline{\tilde{x}} = \underline{\underline{F}}_Q \cdot \underline{d} \quad (4.46)$$

At the receiver, an IFFT is performed after the channel equalization in the frequency domain to invert the FFT linear pre-coder:

$$\underline{\hat{d}} = \underline{\underline{F}}_Q^H \cdot \underline{\underline{\Lambda}}_E \cdot \underline{\tilde{y}} \quad (4.47)$$

Note that we have again omitted the block index in the two last expressions for the sake of clarity. As a result, the signal is transmitted in the time domain, convolved with the time dispersive channel, converted at the receiver to the frequency domain for low complexity channel equalization, and converted back to the time domain for symbol detection. By combining Equations (4.24), (4.46) and (4.47), we obtain a matrix model for the SC-FDE system in the presence of CFO, SCO and IQ imbalance:

$$\underline{\tilde{d}} = \underline{\underline{F}}_Q^H \cdot \underline{\underline{\Lambda}}_E \cdot (\alpha \underline{\gamma} \cdot \underline{\underline{\Lambda}}_0 \cdot \underline{\underline{\Lambda}}_{\tilde{h}} \cdot (\underline{\underline{F}}_Q \cdot \underline{d}) + \beta \underline{\tilde{\gamma}}^* \cdot \underline{\underline{\Lambda}}_0^* \cdot \underline{\underline{\Lambda}}_{\tilde{h}}^* \cdot (\underline{\underline{F}}_Q^* \cdot \underline{d}^*) + \underline{\tilde{z}}) \quad (4.48)$$

The definitions (4.21), (4.22) and (4.23) of the constituting terms γ_0^q (phase and modulus) and $\gamma^{p,q}$ of the matrices $\underline{\underline{\Lambda}}_0$ and $\underline{\underline{\gamma}}$ still hold.

Even if the SC-FDE system can be seen as a special case of the OFDM system, it has a significantly different behavior in the presence of front-end non-idealities. In the case of the AWGN channel, the SC-FDE receiver reduces to gathering the received samples in a vector (the equalizer is a scaled identity matrix so that the IFFT annihilates the FFT). Therefore, the front-end non-idealities are directly seen in the time domain. In other words, CFO will cause a phase ramp on the received vector, SCO will cause a time interpolation of the received samples and IQ imbalance will superpose the received signal and its conjugate. In the case of low-delay spread time dispersive channels, the behavior will be similar to the AWGN channel case. In the case of high-delay spread time dispersive channels, the effect of the non-idealities will be spread over the neighboring symbols.

In the following sections, we particularize the model (4.48) for each non-ideality independently. We rely on the approximated OFDM model (4.34) better to understand the effect of each non-ideality.

4.2.1 Effect of carrier frequency offset

When the system suffers only from CFO (no SCO and no IQ imbalance), the expression (4.48) becomes:

$$\hat{\underline{d}} = \underline{\underline{F}}_Q^H \cdot \underline{\underline{\Lambda}}_E \cdot \underline{\underline{\gamma}} \cdot \underline{\underline{\Lambda}}_0 \cdot \underline{\underline{\Lambda}}_{\tilde{h}} \cdot \underline{\underline{F}}_Q \cdot \underline{d} + \underline{\underline{F}}_Q^H \cdot \underline{\underline{\Lambda}}_E \cdot \underline{\tilde{z}} \quad (4.49)$$

where the constituting terms γ_0^q and $\gamma^{p,q}$ of the matrices $\underline{\underline{\Lambda}}_0$ and $\underline{\underline{\gamma}}$ are given in Equations (4.26), (4.27) and (4.28). The expression (4.49) is approximately equal to (see Equation (4.34)):

$$\hat{\underline{d}} = e^{-j\phi_0}(\underline{\underline{F}}_Q^H \cdot \underline{\underline{\Lambda}}_E \cdot \underline{\underline{\gamma}} \cdot \underline{\underline{\Lambda}}_{\tilde{h}} \cdot \underline{\underline{F}}_Q) \cdot \underline{d} + \underline{\underline{F}}_Q^H \cdot \underline{\underline{\Lambda}}_E \cdot \underline{\tilde{z}} \quad (4.50)$$

where the terms $\gamma^{p,q}$ are given by:

$$\gamma^{p,q} := \frac{(-1)^{(q-p)} \cdot e^{-j\pi(q-p)/Q}}{\cos(\frac{\pi(q-p)}{Q}) - (\frac{\Delta\omega T}{2})^{-1} \sin(\frac{\pi(q-p)}{Q})} \quad (4.51)$$

Therefore, CFO causes a common phase rotation equal to the one caused on the OFDM system and interference of each symbol on its the neighboring symbols in the SC-FDE block. In the case of the AWGN channel, the expression (4.50) further reduces to:

$$\hat{\underline{d}} = e^{-j\phi_0}\underline{\underline{\Lambda}}_\gamma \cdot \underline{d} + \underline{\underline{F}}_Q^H \cdot \underline{\tilde{z}} \quad (4.52)$$

in which, based on the properties of circulant matrices, it can be seen that $\underline{\underline{\Lambda}}_\gamma := \underline{\underline{F}}_Q^H \cdot \underline{\underline{\gamma}} \cdot \underline{\underline{F}}_Q$ is a diagonal matrix composed of the phase shifts $e^{j\Delta\omega nT}$ on its diagonal ($n = -Q/2, \ldots, Q/2 - 1$). Therefore, the interference between the symbols caused by the CFO is canceled out in the case of the AWGN channel (contrary to the OFDM system).

Figure 4.15 illustrates the performance of the SC-FDE system in the presence of CFO for the AWGN channel. We evaluate the system performance when the phase ramp generated

IMPACT OF THE NON-IDEAL FRONT-ENDS

by the CFO is or is not compensated. When the phase ramp is compensated, the effect of the CFO is entirely removed from the received signal so that the system achieves the performance of the ideal system. When the phase ramp is not compensated, the performance degradation due to the CFO is comparable for the OFDM and SC-FDE systems (compare the left part of Figure 4.5 to Figure 4.15).

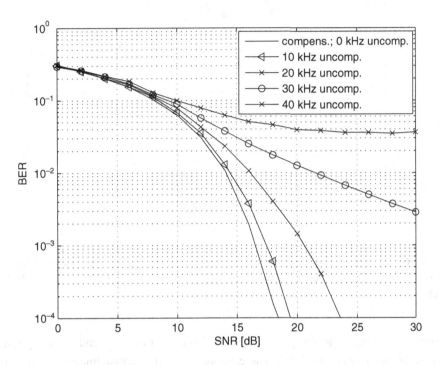

Figure 4.15 Impact of the CFO on the system performance in the case of the AWGN channel.

Figure 4.16 illustrates the performance of the SC-FDE system in the presence of CFO for time-dispersive channels. The case of low-delay spread channels (solid curves) is compared to the case of high-delay spread channels (dashed curves). In the case of time-dispersive channels, the phase ramp generated by the CFO results in interference between the neighboring symbols. The CFO can therefore not be compensated (except for the common phase rotation) and the performance of the SC-FDE system in the presence of CFO is close to the performance of the OFDM system (compare Figures 4.6 and 4.16).

4.2.2 Effect of sample clock offset

When the system suffers only from SCO (no CFO and no IQ imbalance), the expression (4.48) becomes:

$$\underline{\hat{d}} = \underline{F}_Q^H \cdot \underline{\underline{\Lambda}}_E \cdot \underline{\gamma} \cdot \underline{\underline{\Lambda}}_0 \cdot \underline{\underline{\Lambda}}_{\tilde{h}} \cdot \underline{F}_Q \cdot \underline{d} + \underline{F}_Q^H \cdot \underline{\underline{\Lambda}}_E \cdot \underline{\tilde{z}} \qquad (4.53)$$

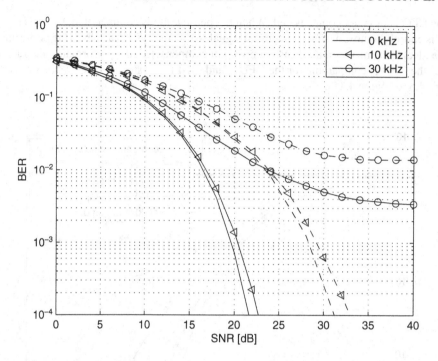

Figure 4.16 Impact of the CFO on the system performance in the case of frequency selective channels (solid curves: low-delay spread channels; dashed curves: high-delay spread channels).

where the constituting terms γ_0^q and $\gamma^{p,q}$ of the matrices $\underline{\underline{\Lambda}}_0$ and $\underline{\underline{\gamma}}$ are given in Equations (4.30), (4.31) and (4.32). The expression (4.53) is approximately equal to (see Equation (4.34)):

$$\hat{\underline{d}} = \underline{\underline{F}}_Q^H \cdot \underline{\underline{\Lambda}}_E \cdot \underline{\underline{\Lambda}}_0 \cdot \underline{\underline{\Lambda}}_{\tilde{h}} \cdot \underline{\underline{F}}_Q \cdot \underline{d} + \underline{\underline{F}}_Q^H \cdot \underline{\underline{\Lambda}}_E \cdot \underline{\tilde{z}} \qquad (4.54)$$

where the terms γ_0^q have a modulus 1 and a phase given by:

$$\angle(\gamma_0^q) := \frac{2\pi t_0}{QT} q \qquad (4.55)$$

Therefore, SCO causes interference between the neighboring symbols of the SC-FDE block. In the case of the AWGN channel, the expression (4.54) further reduces to:

$$\hat{\underline{d}} = (\underline{\underline{F}}_Q^H \cdot \underline{\underline{\Lambda}}_0 \cdot \underline{\underline{F}}_Q) \cdot \underline{d} + \underline{\underline{F}}_Q^H \cdot \underline{\tilde{z}} \qquad (4.56)$$

in which $\underline{\underline{F}}_Q^H \cdot \underline{\underline{\Lambda}}_0 \cdot \underline{\underline{F}}_Q$ is a circulant matrix whose first column is the IFFT of the diagonal of $\underline{\underline{\Lambda}}_0$, equal to a time interpolation filter. Therefore, SCO causes a non-negligible interference of each symbol on its neighboring symbols (contrary to the OFDM system).

Figure 4.17 illustrates the performance of the SC-FDE system in the presence of SCO for the AWGN channel. Compared to the OFDM system for which most of the SCO can be compensated, SCO causes interference between the neighboring symbols in the case of the SC-FDE system that can degrade significantly the performance. The level of interference increases proportionally with the symbol block index so that the number of blocks in the burst should be limited to a reasonable value.

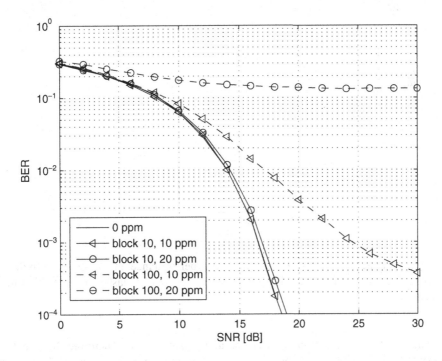

Figure 4.17 Impact of the SCO on the system performance in the case of the AWGN channel.

Figure 4.18 extends the result to time-dispersive channels focusing on block 50 in the burst. Both low-delay spread channels (solid curves) and high-delay spread channels (dashed curves) are illustrated. Compared to the OFDM system, the performance of the SC-FDE system is greatly affected by the SCO. The SCO should be limited to 10 ppm to enable the system to work. Note, however, that the first blocks in the burst suffer less from SCO so that the specification can be relaxed for shorter bursts.

4.2.3 Effect of IQ imbalance

When the system suffers only from IQ imbalance (no CFO and no SCO), the expression (4.48) becomes:

$$\underline{\hat{d}} = \alpha \underline{\underline{F}}_Q^H \cdot \underline{\underline{\Lambda}}_E \cdot \underline{\underline{\Lambda}}_{\tilde{h}} \cdot \underline{\underline{F}}_Q \cdot \underline{d} + \beta \underline{\underline{F}}_Q^H \cdot \underline{\underline{\Lambda}}_E \cdot \underline{\tilde{I}}_Q \cdot \underline{\underline{\Lambda}}_{\tilde{h}}^* \cdot \underline{\underline{F}}_Q^* \cdot \underline{d}^* + \underline{\underline{F}}_Q^H \cdot \underline{\underline{\Lambda}}_E \cdot \underline{\tilde{z}} \quad (4.57)$$

where the factor α is approximately equal to 1 and can be neglected. The matrix $\underline{\tilde{I}}_Q$ is a mirrored identity matrix equal to $\underline{\underline{F}}_Q \cdot \underline{\underline{F}}_Q$ so that the term in β is a circulant matrix

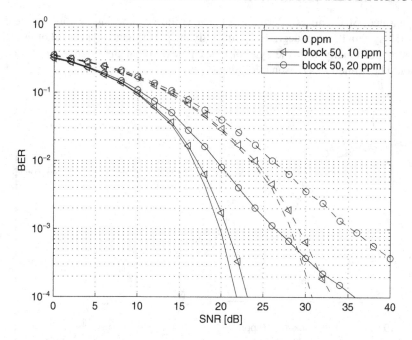

Figure 4.18 Impact of the SCO on the system performance in the case of frequency selective channels (solid curves: low-delay spread channels; dashed curves: high-delay spread channels).

(the product of two circulant matrices is also a circulant matrix). Therefore, IQ imbalance causes interference of each symbol on the neighboring symbols of the SC-FDE block. In the case of the AWGN channel, the expression (4.57) reduces approximately to (see Equation (4.34)):

$$\hat{\underline{d}} = \underline{d} + \beta \underline{d}^* + \underline{\underline{F}}_Q^H \cdot \underline{\tilde{z}} \qquad (4.58)$$

so that IQ imbalance superposes the complex conjugate of the symbol vector to itself. Therefore, IQ imbalance does not cause interference between the symbols in case of the AWGN channel (as opposed to the OFDM system).

Figure 4.19 illustrates the performance of the SC-FDE system in the presence of IQ imbalance for the AWGN channel. We evaluate the system performance when the second term generated by the IQ imbalance is or is not compensated (see Section 5.9.7 on the compensation of the IQ imbalance based on the knowledge of β). After compensation, the IQ imbalance is entirely removed from the received signal so that the system achieves the performance of the ideal system. When the IQ imbalance is not compensated, the performance degradation due to the IQ imbalance is comparable for the OFDM and SC-FDE systems (compare the left part of Figure 4.11 to Figure 4.19).

Figure 4.20 illustrates the performance of the SC-FDE system in the presence of IQ imbalance for time-dispersive channels. The case of low-delay spread channels (solid curves) is compared to the case of high-delay spread channels (dashed curves). In the case of time dispersive channels, the second term generated by the IQ imbalance results in interference

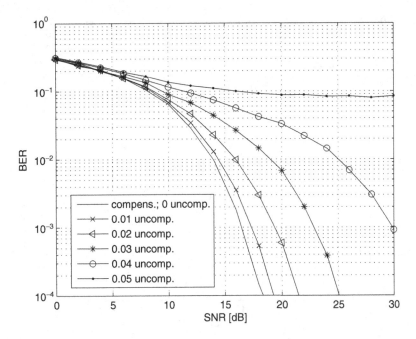

Figure 4.19 Impact of the phase mismatch on the system performance in the case of the AWGN channel.

between the neighboring symbols. The IQ imbalance cannot therefore be compensated and the performance of the SC-FDE system in the presence of IQ imbalance is close to (actually slightly better than) the performance of the OFDM system (compare Figures 4.12 and 4.20).

4.2.4 Combination of effects

We evaluate finally in Figure 4.21 the performance of the SC-FDE system when the non-idealities are present at their recommended specification (10 kHz CFO, 10 ppm SCO and 0.1% amplitude and phase mismatch). We illustrate only the performance of the block 50 in the burst, so that our result is pessimistic compared to the performance averaged over all blocks in a burst of 50 blocks because the effect of SCO on the previous blocks is smaller. Again, the performance is evaluated for low-delay spread channels and for high-delay spread channels separately. Estimation/compensation algorithms must be developed to meet the specifications.

4.3 Comparison of the sensitivity of OFDM and SC-FDE to CFO, SCO and IQ imbalance

To conclude the discussion, we compare the OFDM and SC-FDE system sensitivity to CFO, SCO and IQ imbalance. We consider successively the AWGN and frequency selective channel cases. Table 4.1 summarizes the effects of each non-ideality for the AWGN channel

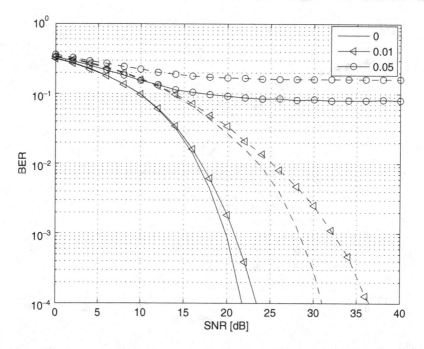

Figure 4.20 Impact of the phase mismatch on the system performance in the case of frequency selective channels (solid curves: low-delay spread channels; dashed curves: high-delay spread channels).

case. Only the ICI/ISI generated by the non-idealities cannot be compensated with reasonably low complexity solutions. It is interesting to observe that CFO and SCO have a dual effect: CFO causes ICI on the OFDM system and a phase ramp on the SC-FDE system while SCO causes a phase ramp on the OFDM system and ISI on the SC-FDE system. The SC-FDE system has the advantage over the OFDM system that the IQ imbalance can be eliminated.

Table 4.1 Comparison of OFDM and SC-FDE in the case of the AWGN channel.

	OFDM	SC-FDE
CFO	Common phase rotation	Common phase rotation
	ICI on neighboring carriers	Phase ramp
SCO	Phase ramp	ISI on neighboring symbols
IQ imbalance	ICI on mirror carrier	Complex conjugated symbol addition

Table 4.2 summarizes the effects of each non-ideality for the time-dispersive channel case. Compared to the AWGN channel case, the effects of the non-idealities on the OFDM system are similar. Only the IQ imbalance has a more pronounced impact because a high-gain carrier can interfere on a low-gain carrier. Unfortunately the SC-FDE system loses its

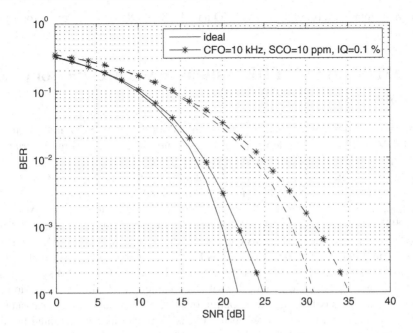

Figure 4.21 System performance in the presence of 10 kHz CFO, 10 ppm SCO and 0.1% amplitude and phase mismatch (solid curves: low-delay spread channels; dashed curves: high-delay spread channels).

Table 4.2 Comparison of OFDM and SC-FDE in the case of time-dispersive channels.

	OFDM	SC-FDE
CFO	Common phase rotation	Common phase rotation
	ICI on neighboring carriers	ISI on neighboring symbols
SCO	Phase ramp	ISI on neighboring symbols
IQ imbalance	ICI on mirror carrier	ISI on neighboring symbols

Table 4.3 Resulting specifications.

	OFDM	SC-FDE
CFO	10 kHz	No limit (AWGN), 10 kHz (multi-path)
SCO	No limit	10 ppm
IQ imbalance	1% (AWGN), $\ll$ 1% (multi-path)	No limit (AWGN); < 1% (multi-path)

nice non-ideality compensation ability compared to the AWGN channel case. Both CFO and IQ imbalance generate ISI when the channels are time-dispersive.

The resulting non-ideality specifications are given in Table 4.3. The difference is made between the AWGN and the time-dispersive channels. In the case of the AWGN channel, the SC-FDE system is, on the whole, more robust to the front-end non-ideality than the OFDM system. In the case of time-dispersive channels, each system has its own advantages:

the OFDM system is more robust to the SCO and the SC-FDE system is more robust to the IQ imbalance.

4.4 OFDM and SC-FDE systems in the presence of phase noise

In the previous sections, we have analyzed the impact of the CFO, SCO and IQ imbalance on the OFDM and SC-FDE systems. Because those non-idealities are time invariant, they can be reliably estimated and removed from the received signal. Another non-ideality of particular interest is the phase noise (PN) generated by the local oscillators. Because the PN is correlated over time, it can also be partially compensated.

The impact of PN on the OFDM system has been initially analyzed by Armada and Calvo (1998). It has been shown that PN causes a common phase rotation and ICI within the OFDM symbol block. A comparison of the sensitivity to PN of the OFDM and single-carrier with time-domain equalization systems has been carried out for different scenarios. In particular, Pollet *et al.* (1995) show that the BER is higher for the OFDM system than for the single-carrier system in the case of the AWGN channel. To obtain the expression of the BER, the authors approximate the contribution of the PN by an additional Gaussian noise component, which is only accurate when the ICI is dominant as demonstrated by Tomba (1998). On the other hand, it has been shown by Moeneclaey (1997) that the ratio of the signal power to interference power is the same for both air interfaces if the common phase rotation is not corrected. Finally Zamorano *et al.* (2007) have extended the comparison to SC-FDE assuming both AWGN and multi-path channels.

4.4.1 System model

A block diagram of the linearly pre-coded OFDM system in the presence of PN is provided in Figure 4.22. The linear pre-coding matrix can be defined such that uncoded OFDM and SC-FDE are instantiated:

- OFDM is obtained by setting $\underline{\underline{\Theta}} = \underline{\underline{I}}_Q$;
- SC-FDE is obtained by setting $\underline{\underline{\Theta}} = \underline{\underline{F}}_Q$.

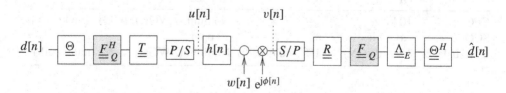

Figure 4.22 The OFDM system in the presence of phase noise $\phi[n]$.

The received sequence, obtained by convolving the transmitted sequence, $u[n]$, with the digital equivalent channel impulse response, $h[n]$, is corrupted by the addition of white

IMPACT OF THE NON-IDEAL FRONT-ENDS

Gaussian noise, $w[n]$, and by a rotation due to the PN with a random phase, $\phi[n]$:

$$v[n] = e^{j\phi[n]} (u[n] \star h[n] + w[n]) \tag{4.59}$$

The random phase $\phi[n]$ is usually correlated over the time samples (see the power frequency components due to the reference crystal noise and to the VCO noise illustrated in Figure 3.19). Taking the PN into account, the estimated symbol vector can be expressed as:

$$\hat{\underline{d}} = \underline{\underline{\Theta}}^H \cdot \underline{\underline{\Lambda}}_E \cdot \underline{\underline{F}}_Q \cdot \underline{\underline{\Phi}} \cdot (\underline{\underline{\dot{H}}} \cdot \underline{\underline{F}}_Q^H \cdot \underline{\underline{\Theta}} \cdot \underline{d} + \underline{z}) \tag{4.60}$$

$$= \underline{\underline{\Theta}}^H \cdot \underline{\underline{\Lambda}}_E \cdot \underline{\underline{F}}_Q \cdot \underline{\underline{\Phi}} \cdot \underline{\underline{F}}_Q^H \cdot \underline{\underline{\Lambda}}_{\tilde{h}} \cdot \underline{\underline{\Theta}} \cdot \underline{d} + \underline{\underline{\Theta}}^H \cdot \underline{\underline{\Lambda}}_E \cdot \underline{\underline{F}}_Q \cdot \underline{\underline{\Phi}} \cdot \underline{z} \tag{4.61}$$

where $\underline{\underline{\Phi}}$ is a diagonal matrix of size Q composed of the random phase rotation of the block samples on its diagonal.

4.4.2 Impact of the PN

It is well known that PN causes a common phase rotation and ICI in the case of OFDM. In this section, we generalize this result for both OFDM and SC-FDE.

In order to highlight the common phase rotation for both air interfaces, the PN diagonal matrix $\underline{\underline{\Phi}}$ can be seen as the addition of the average of the PN elements $\bar{\Phi} := \frac{1}{Q}\text{Tr}(\underline{\underline{\Phi}})$ plus a deviation diagonal matrix $\underline{\underline{\Delta}}_\Phi$:

$$\underline{\underline{\Phi}} = \bar{\Phi}\underline{\underline{I}}_Q + \underline{\underline{\Delta}}_\Phi \tag{4.62}$$

By substituting Equation (4.62) in Equation (4.61), we obtain:

$$\hat{\underline{d}} = \bar{\Phi}\underline{\underline{\Theta}}^H \cdot \underline{\underline{\Lambda}}_E \cdot \underline{\underline{\Lambda}}_{\tilde{h}} \cdot \underline{\underline{\Theta}} \cdot \underline{d}$$
$$+ \underline{\underline{\Theta}}^H \cdot \underline{\underline{\Lambda}}_E \cdot \underline{\underline{F}}_Q \cdot \underline{\underline{\Delta}}_\Phi \cdot \underline{\underline{F}}_Q^H \cdot \underline{\underline{\Lambda}}_{\tilde{h}} \cdot \underline{\underline{\Theta}} \cdot \underline{d} + \underline{\underline{\Theta}}^H \cdot \underline{\underline{\Lambda}}_E \cdot \underline{\underline{F}}_Q \cdot \underline{\underline{\Phi}} \cdot \underline{z} \tag{4.63}$$

The first term is the estimated symbol block in the absence of PN multiplied by the factor $\bar{\Phi}$, approximately equal to the phase rotation averaged over the block. For small values of the phase errors, it is indeed approximately equal to:

$$\bar{\Phi} = \frac{1}{Q}\sum_{n=0}^{Q-1} e^{j\phi[n]} \tag{4.64}$$

$$\approx 1 + \frac{j}{Q}\sum_{n=0}^{Q-1} \phi[n] \tag{4.65}$$

$$\approx e^{j1/Q \sum_{n=0}^{Q-1} \phi[n]} \tag{4.66}$$

in which the first-order approximation $e^{j\phi[n]} \approx 1 + j\phi[n]$ for small values of the phase error ($|\phi[n]| \ll 1$) has been used. The second term corresponds to the interference between the symbols of the block (ICI in the case of OFDM, ISI in the case of SC-FDE). It is useful to develop the inner matrix product as:

$$\underline{\underline{F}}_Q \cdot \underline{\underline{\Delta}}_\Phi \cdot \underline{\underline{F}}_Q^H = \underline{\underline{F}}_Q^H \cdot \underline{\underline{P}} \cdot \underline{\underline{\Delta}}_\Phi \cdot \underline{\underline{P}}^H \cdot \underline{\underline{F}}_Q \tag{4.67}$$

$$= \underline{\underline{F}}_Q^H \cdot \underline{\underline{\Delta}}'_\Phi \cdot \underline{\underline{F}}_Q \tag{4.68}$$

in which $\underline{\underline{P}}$ is the permutation matrix equal to $\underline{\underline{P}} := \underline{\underline{F}}_Q \cdot \underline{\underline{F}}_Q$ and $\underline{\underline{\Delta}}'_\Phi := \underline{\underline{P}} \cdot \underline{\underline{\Delta}}_\Phi \cdot \underline{\underline{P}}^H$ is a diagonal matrix composed of the diagonal elements of $\underline{\underline{\Delta}}_\Phi$ in reverse order.

In the case of the AWGN channel, the equality (4.63) reduces to:

$$\hat{\underline{d}} = \bar{\Phi} \cdot \underline{d} + \underline{\underline{F}}_Q^H \cdot \underline{\underline{\Delta}}'_\Phi \cdot \underline{\underline{F}}_Q \cdot \underline{d} + \underline{\underline{F}}_Q \cdot \underline{\underline{\Phi}} \cdot \underline{z} \qquad (4.69)$$

for OFDM, and to:

$$\hat{\underline{d}} = \bar{\Phi} \cdot \underline{d} + \underline{\underline{\Delta}}_\Phi \cdot \underline{d} + \underline{\underline{F}}_Q \cdot \underline{\underline{\Phi}} \cdot \underline{z} \qquad (4.70)$$

for SC-FDE. There is ICI in the case of OFDM because the matrix $\underline{\underline{F}}_Q^H \cdot \underline{\underline{\Delta}}'_\Phi \cdot \underline{\underline{F}}_Q$ is circulant (its diagonal is composed of zeros, since the average of the PN elements is included in the first term of Equation (4.69)). The PN results in clouds around the desired constellation points due to the ICI. On the contrary, there is no ISI in the case of SC-FDE because the matrix $\underline{\underline{\Delta}}_\Phi$ is diagonal. The PN results in arcs around the desired constellation points corresponding to random rotations of the constellation points.

In the case of time-dispersive channels, there are both ICI in the OFDM system and ISI in the SC-FDE system. It generates a dispersion of the received constellation in both cases. Even if the average power of the ICI and ISI is equal (this is easily shown by computing the trace of the auto-correlation of the second term for each air interface), the received constellation points can be more distant from the initial constellation points in the SC-FDE case than in the OFDM case because they correspond to a dispersion of an arc instead of a circle.

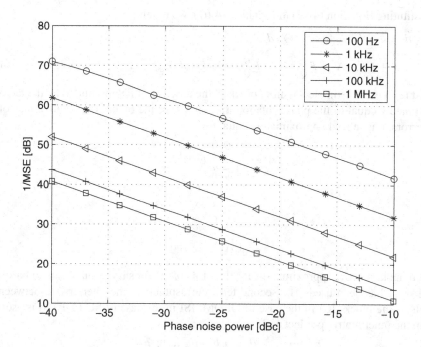

Figure 4.23 Impact of the PN power on the MSE for a varying PN cutoff frequency in the case of the AWGN channel.

Consequently, SC-FDE can be slightly more sensitive to PN as will be confirmed in the BER simulations.

4.4.3 Numerical analysis

Figure 4.23 illustrates the SNR, defined in this study as the inverse of the mean square error (MSE), after compensation of the common phase rotation for the AWGN channel. Because the PN generates random phase components causing a performance degradation, the MSE is better suited than the SIR to evaluate the impact of the PN. The MSE is identical for the OFDM and SC-FDE systems because the common phase rotation and the average power of the ICI and ISI are the same. The impact of the PN power and cutoff frequency is assessed. If the cutoff frequency is smaller than the sub-carrier spacing (20 MHz /64 $\approx$ 300 kHz), the dominant effect is the common phase rotation that can be compensated. If the cutoff frequency is larger than the sub-carrier spacing, the common phase rotation becomes negligible compared to the ICI and ISI. The resulting SNR (defined as 1/MSE) is approximately equal to the PN power in the last case.

Figure 4.24 illustrates the performance of the OFDM and SC-FDE systems in the presence of PN for the AWGN channel. The impact of the PN cutoff frequency is evaluated in the left part of Figure 4.24 for the OFDM system assuming that the common phase rotation is corrected. When the PN cutoff frequency equals 10 kHz, the PN is nearly entirely corrected

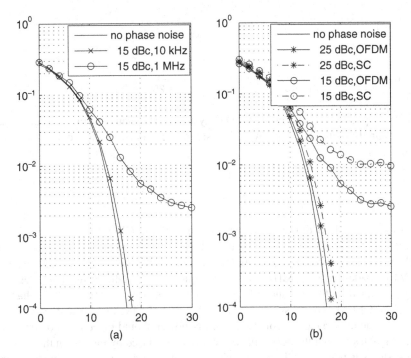

Figure 4.24 Impact of the PN on the system performance in the case of the AWGN channel; (a) OFDM system, -15 dBc PN power, varying PN cutoff frequency; (b) OFDM versus SC-FDE system, varying PN power, 1 MHz PN cutoff frequency.

so that the performance degradation is negligible. When the PN cutoff frequency equals 1 MHz, the PN cannot be corrected so that the performance degradation is high. The same conclusion can be drawn for the SC-FDE system. The right part of Figure 4.24 demonstrates that the OFDM system is slightly more robust than the SC-FDE system to PN when the ICI/ISI is dominant.

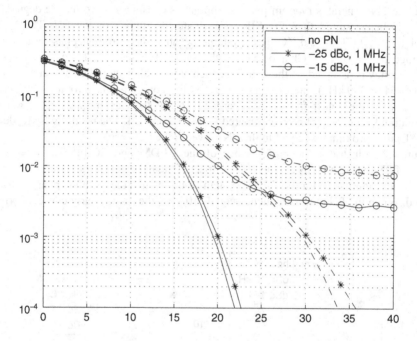

Figure 4.25 Impact of the PN on the OFDM system performance in the case of frequency selective channels (solid curves: low-delay spread channels; dashed curves: high-delay spread channels).

Figures 4.25 and 4.26 illustrate the performance of the OFDM and SC-FDE systems, respectively, in the presence of PN for time-dispersive channels. The case of low-delay spread channels (solid curves) is compared to the case of high-delay spread channels (dashed curves). We consider again a 1 MHz PN cutoff frequency so that the performance degradation is mainly generated by the ICI and ISI. The conclusions drawn for the AWGN channel are still valid. The OFDM system is more robust to PN than the SC-FDE system even when the channels are time dispersive. To conclude, the PN power should be equal to or smaller than -25 dBc to make sure that the performance degradation caused by the PN on the OFDM and SC-FDE systems is negligible. However, when the inter-carrier spacing is much larger than the PN cutoff frequency, the PN can be nearly entirely removed from the received signal and the specification on the PN power is relaxed.

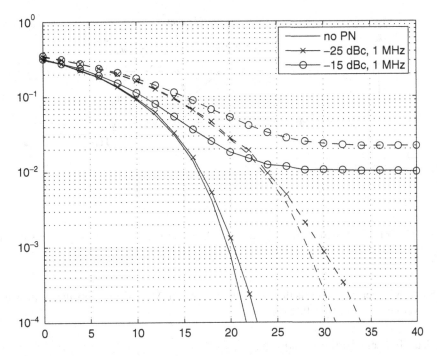

Figure 4.26 Impact of the PN on the SC-FDE system performance in the case of frequency selective channels (solid curves: low-delay spread channels; dashed curves: high-delay spread channels).

4.5 OFDM system in the presence of clipping, quantization and nonlinearity

4.5.1 Clipping

Because of the finite full-scale range of ADCs and DACs, clipping can occur at the interface between the analog and the digital domains. Figure 4.27 shows the voltage transfer function of a simple 3-bit ADC (left) and the output of this ADC for a one-cycle sinusoid (right). As can be seen from the error signal, the quantization noise can be expected to be in first order uniformly distributed whereas the clipping noise distribution is not so easy to determine. Although both phenomena occur simultaneously in the converter, their characterization and effect on the wireless signal is quite different. In addition, clipping can differ according to the TX or RX architecture. Indeed, when quadrature is generated in the analog domain, separate DACs or ADCs are used for I and Q whereas, when quadrature is generated digitally, the I and Q signals are converted by a single DAC or ADC. The net difference is that, in the former case, it is the distribution of $x_I(t)$ or $x_Q(t)$ that determines the clipping probability and in the latter case, it is the distribution of $x_L(t) = \sqrt{x_I^2(t) + x_Q^2(t)}$ that determines the clipping

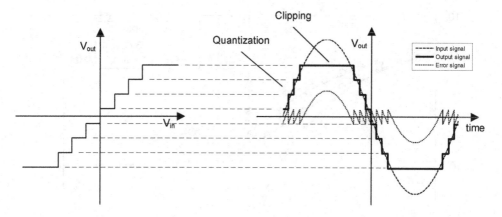

Figure 4.27 Transfer function of an ideal 3-bit ADC (left) and quantization of a sinewave (right).

probability. Another difference is that, for a single DAC or ADC, the mean value of the signal to be converted is higher by a factor $\sqrt{2}$.

Statistical characterization of clipping

A reasonable assumption to characterize an OFDM waveform is to exploit the central limit theorem. By virtue of its definition, we have:

$$x_L(t) = x_I(t) + jx_Q(t) = \frac{1}{\sqrt{Q}} \sum_{q=-Q/2}^{Q/2-1} \tilde{x}^q \, e^{j(2\pi qt/QT)} \quad (4.71)$$

which shows that an OFDM waveform is the sum of a large number (typically 50 to 8000 for modern standards) of weighted sinusoids. Hence, the amplitudes of the I and Q signals $x_I(t)$ and $x_Q(t)$ tend to a Gaussian distribution and the amplitude of $x_L(t)$ tends to a Rayleigh distribution, irrespective of the distribution of the information symbols $\tilde{x}^q$.

We define the variances of $x_I(t)$, $x_Q(t)$ and $x_L(t)$ as σ_I^2, σ_Q^2 and σ_L^2, respectively. We have:

$$\sigma_I = \sigma_Q \quad (4.72)$$

$$\sigma_L = 2\sigma_I \quad (4.73)$$

The signal is clipped by the ADC when its amplitude exceeds the clipping level $\pm A_{CL}$. Given σ_I, we can derive the clip probability P_{clip,x_I} and P_{clip,x_L} of $x_I(t)$ and $x_L(t)$, respectively:

$$P_{clip,x_I} = P(|x_I| > A_{CL}) = 2 \int_{A_{CL}}^{\infty} \frac{1}{\sigma\sqrt{2\pi}} e^{-x^2/2\sigma_{x_I}^2} \, dx = \mathrm{erfc}\left(\frac{A_{CL}}{\sqrt{2}\sigma_I}\right) \quad (4.74)$$

$$P_{clip,x_L} = P(|x_L| > A_{CL}) = \int_{A_{CL}}^{\infty} \frac{x}{\sigma_L^2} e^{-x^2/2\sigma_L^2} \, dx = e^{-A_{CL}^2/2\sigma_L^2} \quad (4.75)$$

IMPACT OF THE NON-IDEAL FRONT-ENDS

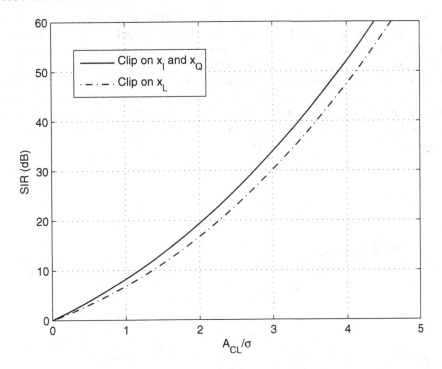

Figure 4.28 SIR degradation of OFDM with clipping. For higher values of the normalized clip level (CL = A_{CL}/σ), the clip probability is reduced and the SIR improves.

The probability P_{clip,x_I} is of interest when the I and Q branches are clipped separately, whereas P_{clip,x_L} applies when the absolute value $|x_L(t)|$ is being clipped. The corresponding clip powers σ^2_{clip,x_I} and σ^2_{clip,x_L} can be derived also:

$$\sigma^2_{clip,x_I} = 2 \int_{A_{CL}}^{\infty} \frac{1}{\sigma\sqrt{2\pi}} (x - A_{CL})^2 e^{-x^2/2\sigma_I^2} \, dx \tag{4.76}$$

$$= \frac{-2A_{CL}\sigma_I}{\sqrt{2\pi}} e^{-A_{CL}^2/2\sigma_I^2} + (A_{CL}^2 + \sigma_I^2) \operatorname{erfc}\left(\frac{A_{CL}}{\sigma_I\sqrt{2}}\right) \tag{4.77}$$

$$\sigma^2_{clip,x_L} = \int_{A_{CL}}^{\infty} \frac{x}{\sigma_L^2} (x - A_{CL})^2 e^{-x^2/2\sigma_L^2} \, dx \tag{4.78}$$

$$= 2\sigma_L^2 e^{-A_{CL}^2/2\sigma_L^2} - A_{CL}\sigma_L\sqrt{2\pi} \operatorname{erfc}\left(\frac{A_{CL}}{\sqrt{2}\sigma_L}\right) \tag{4.79}$$

Performance of OFDM with clipping

The clip power from Equation (4.79) can be used to predict the SIR resulting from clipping (Figure 4.28). The abscissa in Figure 4.28 is the ratio of the clip level A_{CL} to the standard deviation of either x_I or x_Q. Obviously, for higher values of A_{CL}/σ, the clipping probability decreases and the SIR improves. In Figure 4.29, the BER performance of OFDM with clipping (separate clipping on I and Q) in an AWGN channel has been estimated by simulation, for different values of the clip level (CL = A_{CL}/σ). These results apply to transmit or receive clipping in AWGN channels. It is important to realize that the SIR as derived earlier cannot be used directly to predict the BER. The reason for this is twofold: first, the noise resulting from clipping as estimated by Equation (4.79) is not Gaussian distributed; second, the distribution of clipping event in *time* is difficult to predict (see, for example, Ochiai and Imai (2000)) and their occurrence is rare at high clip level, with the result that they do not affect all OFDM symbols in the same way.

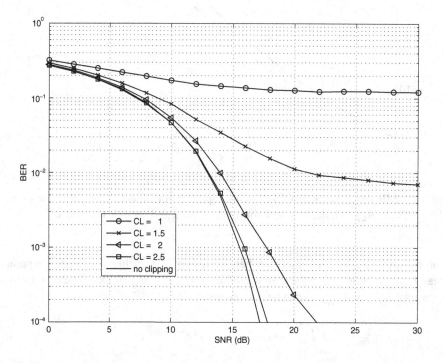

Figure 4.29 Impact of clipping on the OFDM system performance in an AWGN channel. The parameter is the clip level (CL = A_{CL}/σ).

4.5.2 Quantization

For a N_{bit} ADC or DAC with clip level A_{CL}, the quantization step corresponds to the amplitude of the LSB and is given by:

$$A_{LSB} = A_{CL}/2^{N_{bit}-1} \tag{4.80}$$

IMPACT OF THE NON-IDEAL FRONT-ENDS

The quantization noise amplitude can be in first order described by a uniform distribution between $-A_{LSB}/2$ and $A_{LSB}/2$. Its PSD is roughly white.[1] As shown in Rabiner and Gold (1975), the quantization noise variance is given by $\sigma_{Quant}^2 = -A_{LSB}^2/12$. Hence, the PSD is constant and also equal to σ_{Quant}^2 (dBc/Hz). Since the quantization noise depends on the LSB amplitude, it also depends on the full-scale value and the number of bits. As already pointed out in Section 3.3.7, there is, for a given number of bits, an optimum between the degradation caused by clipping and quantization. Based on the results from the previous section, we take a clipping level A_{CL}/σ of 2.5 as acceptable for OFDM in AWGN channels. The resulting BER for OFDM in AWGN with quantization is shown in Figure 4.30. The variance of the quantization noise is also shown in Figure 4.31. These results apply to transmit or receive quantization in AWGN channels.

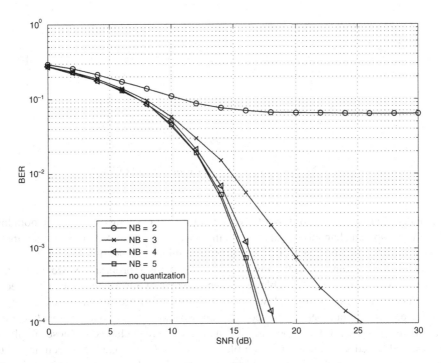

Figure 4.30 Impact of quantization on the OFDM system performance in the case of the AWGN channel ($A_{CL}/\sigma = 2.5$).

4.5.3 Clipping and quantization in frequency selective channels

So far, we have analyzed clipping and quantization in an AWGN channel. The performance in frequency selective fading channels is also important since OFDM is normally used in

---

[1] This is the case for ADC architectures such as flash, successive approximation and pipeline. Sigma-delta converters, by virtue of their noise-shaping properties, do not feature a white PSD.

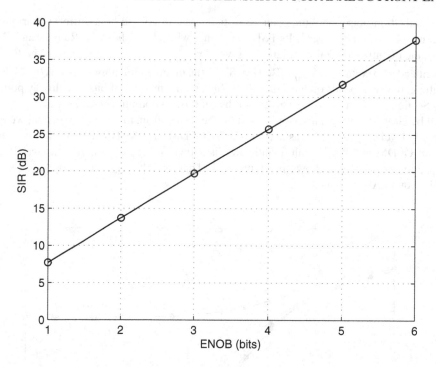

Figure 4.31 SIR versus ENOB for quantized OFDM.

such channels. Figure 4.32 (top) shows the effect of receiver clipping on OFDM in both low- and high-delay spread channels. It can be seen that, in low-delay spread channels, a clipping level of about 2.5 or better is sufficient to guarantee a minor degradation at a BER of 10^{-3}. In high-delay spread channels, a clipping level of 3 or more is appropriate. Since the value of the delay spread is usually not available in advance, a value of 3 is a safe value. The effect of receiver quantization is shown in the lower part of Figure 4.32. Here, the clip level was set at a fixed value of 3 and bit number was varied between two and five. We conclude that five bits is the minimum acceptable value, especially for high-delay spread channels. These simulation results were obtained with the assumption of a perfect AGC. To take the inaccuracy of the AGC into account, it would be safer to add one extra bit of resolution, resulting in six bits for this particular example (one additional bit provides 6 dB of extra dynamic range).

4.5.4 Spectral regrowth with clipping

Besides affecting the BER's 'own' signal, clipping has another adverse effect on the signals in the adjacent channels: spectral regrowth. The nice spectral properties of the OFDM signal are kept only if the transmitter blocks have a linear behavior. Clipping is highly nonlinear and creates significant spectral regrowth. This is illustrated in Figure 4.33: the PSD has been computed by simulation for various levels of clipping. This is clearly a problem that must be tackled in the transmitter chain, since regulatory constraints usually impose tight limits on the amount of out-of-band emission. One obvious way to cope with this is to use an analog

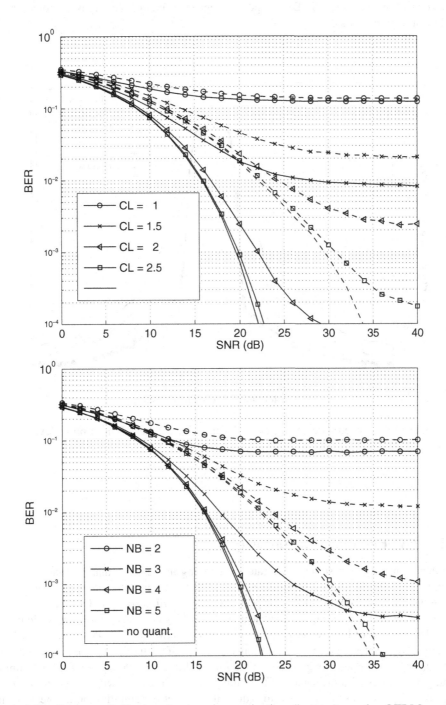

Figure 4.32 Impact of clipping (top) and quantization (bottom) on the OFDM system performance in the case of frequency selective channels (solid curves: low-delay spread channels; dashed curves: high-delay spread channels).

spectral shaping filter after the DAC. Spectral regrowth due to clipping is obviously not a problem at the receiver side.

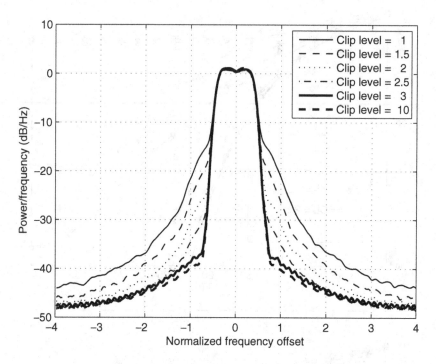

Figure 4.33 Power spectral density of OFDM, for various values of the clipping level (CL = A_{CL}/σ).

4.5.5 Power amplifier nonlinearity

The PA nonlinearity is in some way related to clipping in the sense that clipping is a 'hard limiting' phenomenon whereas PA nonlinearity is a 'soft limiting' phenomenon. Hence, it can be expected that BER degradation and spectral regrowth will also affect OFDM modulation when a nonlinear PA is used. PA nonlinearity is actually the most significant difficulty in the implementation of low-power (i.e. battery-operated terminals) OFDM transmitters. This difficulty is aggravated by technology scaling (reduction of the track widths in semiconductor devices), since this goes along with a reduction of the supply voltage and the maximum output power varies with the square of the supply voltage. Back-off with respect to the 1 dB compression point is usually needed to reduce the impact of saturation on the peak of the OFDM waveform. To illustrate this, we show simulation results with PA nonlinearity in which we have used a third-order PA model, as introduced in Section 3.3.1.

IMPACT OF THE NON-IDEAL FRONT-ENDS

First, we analyze the BER degradation in AWGN channels (Figure 4.34). Several curves are shown with the input back-off as a parameter, varying between +6 and −6 dB.[2] In AWGN channels, 6 dB of back-off provides a satisfying performance.

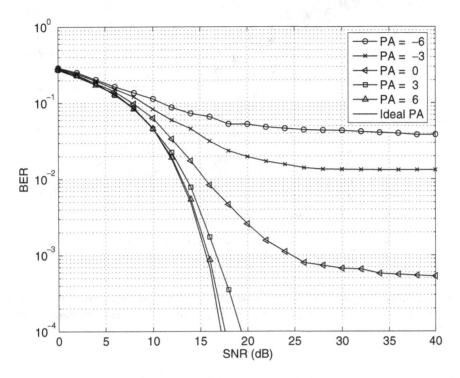

Figure 4.34 Impact of nonlinear PA on the OFDM system performance in the AWGN channel.

In multi-path channels, the effect of saturation is slightly increased. This can be seen in Figure 4.35 that shows the BER degradation due to PA saturation in both low- and high-delay spread channels.

4.5.6 Spectral regrowth with PA nonlinearity

In addition – and just as for clipping – nonlinear amplification also degrades the nice spectral properties of the OFDM signal. Figure 4.36 shows the PSD of an OFDM signal for various levels of back-off, indicating that the spectral regrowth increases as the PA is driven closer to saturation. Unlike clipping, the spectral regrowth at RF is usually impossible to compensate for and this is for two reasons: first, the analog bandpass filter at RF needed to reshape the spectrum is not realizable (the filter Q-factor, determining its selectivity, would be much too

---

[2]The input back-off is the relative difference between the PA input power and the input-referred 1 dB compression point; a negative value for the back-off means that the input power is higher than the 1 dB compression point.

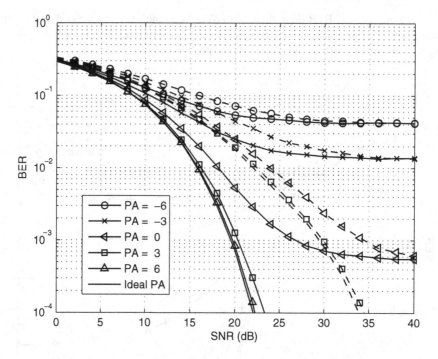

Figure 4.35 Impact of nonlinear amplification on the OFDM system performance in the case of frequency selective channels (solid curves: low-delay spread channels; dashed curves: high-delay spread channels).

high); second, a tunable bandpass filter is needed, since the transmitter can usually transmit at several discrete RF carrier frequencies. Hence, resorting to back-off is the only solution to comply with the regulatory requirements of emission in the adjacent channels. In modern OFDM standards, a back-off of 6 or 7 dB is typical.

4.6 SC-FDE system in the presence of clipping, quantization and nonlinearity

In this section, we want to emphasize the differences and possible advantages of SC-FDE with respect to OFDM when nonlinear elements are present in the communication chain (see Section 4.5). We will first consider receiver clipping and quantization in multi-path channels (we know from the previous study that the effect on low-delay spread channels is lower). Then, we will turn our attention to the transmitter and analyze the PSD degradation due to clipping, quantization and power amplifier nonlinearity.

IMPACT OF THE NON-IDEAL FRONT-ENDS 113

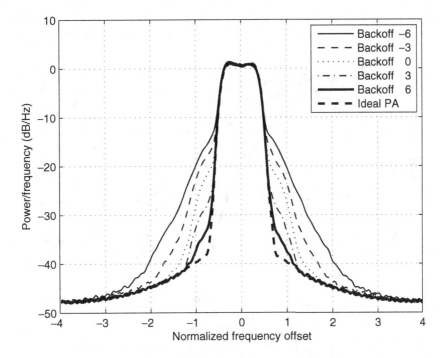

Figure 4.36 PSD of OFDM with PA nonlinearity.

4.6.1 Impact of quantization and clipping at the receiver

Figure 4.37 (top) shows the effect of receiver clipping on SC-FDE in both low- and high-delay spread channels. In low-delay spread channels, a clip level of 2 is adequate whereas, in high delay spread channels, a clip level of 2.5 is required. The effect of quantization is shown in the bottom part of Figure 4.37. Here, the clip level was set at a fixed value of 2.5 and the number of bits was varied between 2 and 5. We observe that 5 bits are needed to guarantee an acceptable BER degradation. Comparing these results with those of Section 4.5.3, we note that SC-FDE does not bring a significant advantage when compared with OFDM. This result may be, at first sight, surprising but it can be explained as follows: although the transmitted SC-FDE waveform has a lower PAPR than the OFDM waveform, the multi-path increases the PAPR of the received signal (because it results in a summation of several delayed replicas of the transmitted waveform). For this reason, the quantization of the received SC-FDE signal affected by multipath must be done with a high clip level (clip level = 2.5) and sufficient resolution (5 bits); these requirements are identical to those that we obtained for the receiver ADC in an OFDM system (Section 4.5.3).

4.6.2 Spectral regrowth with clipping at the transmitter

Thanks to the pulse-shaping filter, SC-FDE can have a very clean output spectrum with out-of-band emissions more than 40 dB below the in-band power. We saw in Section 4.5.4 that clipping has an adverse effect on the signals in the adjacent channels because of spectral

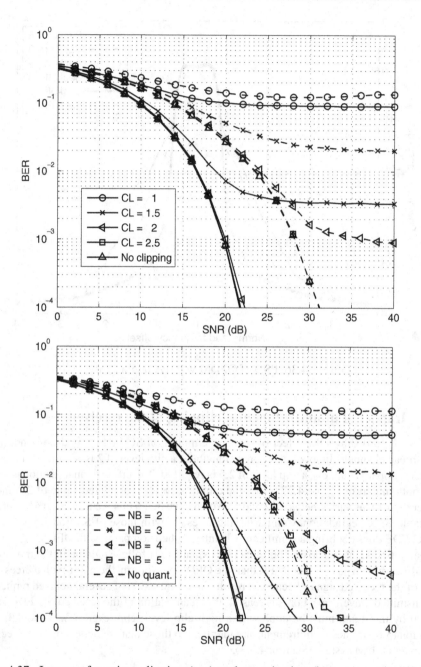

Figure 4.37 Impact of receiver clipping (top) and quantization (bottom) on the SC-FDE system performance in frequency selective channels (solid curves: low-delay spread channels; dashed curves: high-delay spread channels). The top curves provide the clip level requirement (CL = 2.5), which is used in the simulation of quantization.

IMPACT OF THE NON-IDEAL FRONT-ENDS

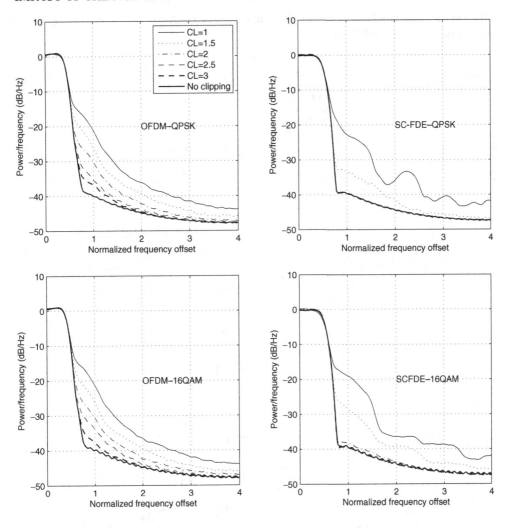

Figure 4.38 Impact of clipping on the PSD of OFDM and SC-FDE, for two different constellations.

regrowth. It is interesting to compare the behavior of OFDM and SC-FDE with respect to transmitter clipping and spectral regrowth. This comparison is illustrated in Figure 4.38 which shows the PSD of OFDM (left plots) and SC-FDE (right plots) for clip levels varying from 1 to 2.5 and without clipping. We also compare the PSD for two different constellation mappings: quadrature phase shift keying (QPSK) and 16QAM. Two main conclusions can be drawn: first, the PSD of OFDM is rather independent from the modulation, because the PSD is dominated by the PAPR, which results in a Gaussian amplitude distribution in the I and Q branches (see Section 4.5.1); second, the PSD of SC-FDE improves rapidly for higher clip levels (a clip level of 1.5 is acceptable and a clip level of 2 generates no degradation in the

PSD. On the other hand, the PSD of OFDM improves more slowly for increasing clip level because of the higher PAPR of OFDM.

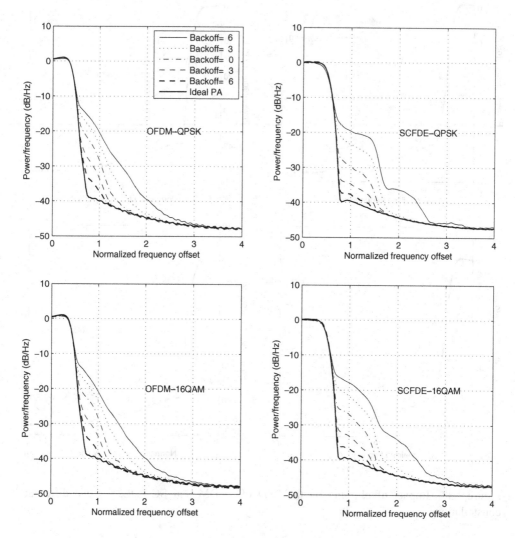

Figure 4.39 PSD of SC-FDE with clipping.

4.6.3 Spectral regrowth with nonlinear PA

The comparison of OFDM and SC-FDE is also interesting in the case of nonlinear PA. Figure 4.39 shows the PSD of OFDM (left plots) and SC-FDE (right plots) for back-off levels (with respect to the 1 dB compression point) varying from -6 dB to $+6$ dB and without nonlinearity. At very low level of back-off (3 or 6 dB above the 1 dB compression point),

IMPACT OF THE NON-IDEAL FRONT-ENDS   117

both systems perform poorly and generate a high level of adjacent channel interference. At 0 dB of back-off, SC-FDE has an acceptable level of adjacent channel interference. Beyond 0 dB of back-off, SC-FDE is very good. At 0 dB of back-off, OFDM generates significant interference close to the desired spectrum. This is undesirable because OFDM channels are usually close to each other because the OFDM spectra have near rectangular PSDs (in the ideal case). Bad performance on sub-carriers close to an adjacent OFDM spectrum will result unless 6 dB of back-off or more is used.

Table 4.4 summarizes the effects of each nonlinearity on OFDM and SC-FDE. RX clipping and quantization produces similar degradation. At the TX side, there is an advantage for SC-FDE that tolerates a lower clip level and needs less PA back-off.

Table 4.4 Comparison of OFDM and SC-FDE in the presence of nonlinearities.

	OFDM	SC-FDE
RX clipping and quantization	CL = 2.5, 5 bits Similar degradation	CL = 2.5, 5 bits Similar degradation
TX clipping and quantization	Slow improvement for increasing clip level	Fast improvement for increasing clip level
PA nonlinearity	Adjacent channel interference needs $\sim$6 dB back-off	Reduced adjacent channel interference needs $\sim$0 dB back-off

4.7 MIMO systems

In this section, we address the impact of non-idealities on MIMO systems. Many non-idealities that we addressed in the previous sections also apply to MIMO systems. Hence, in this section, we will focus on those that are specific to MIMO transmission. A block diagram of a MIMO–OFDM system is shown in Figure 4.40, highlighting several of the non-idealities covered in this section, namely CFO, SCO and antenna mismatch. We will focus our analysis on the three following topics:

- an extension of the model of joint CFO and SCO to MIMO–OFDM systems;
- a comparison of diversity techniques (maximum ratio combining (MRC), STBC) in the presence of non-idealities;
- the impact of non-reciprocity on closed-loop SDM and downlink SDMA systems.

4.7.1 Impact of CFO and SCO on MIMO-OFDM

The impact of CFO and SCO can be analyzed by means of an extension of the model introduced in Section 4.2. For a MIMO–OFDM system without IQ imbalance, transmitting from N_T transmit antennas to N_R receive antennas, the matricial system model becomes:

$$\tilde{\underline{y}} = \underline{\underline{\Gamma}} \cdot \underline{\underline{\check{\Lambda}}}_0 \cdot \underline{\underline{\check{\Lambda}}}_{\tilde{h}} \cdot \tilde{\underline{x}} + \tilde{\underline{z}} \qquad (4.81)$$

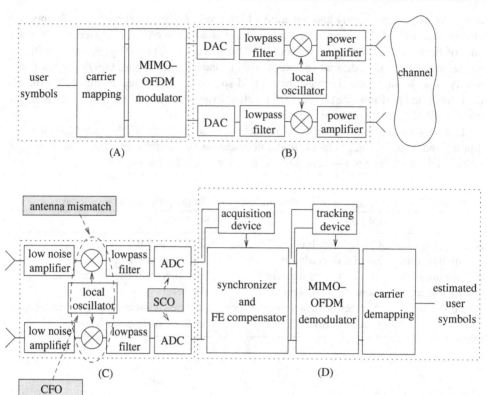

Figure 4.40 MIMO-OFDM system block diagram: (A) transmit digital transceiver; (B) transmit analog front-end; (C) receive analog front-end; (D) receive digital transceiver.

where the constituting terms of the vectors $\underline{\tilde{x}}$, $\underline{\tilde{y}}$ and $\underline{\tilde{z}}$ are defined as:

$$\underline{\tilde{y}} := [\underline{\tilde{y}}_1^T \, \underline{\tilde{y}}_2^T \ldots \underline{\tilde{y}}_{N_R}^T]^T \qquad (QN_R) \times 1 \qquad (4.82)$$

$$\underline{\tilde{x}} := [\underline{\tilde{x}}_1^T \, \underline{\tilde{x}}_2^T \ldots \underline{\tilde{x}}_{N_T}^T]^T \qquad (QN_T) \times 1 \qquad (4.83)$$

$$\underline{\tilde{z}} := [\underline{\tilde{z}}_1^T \, \underline{\tilde{z}}_2^T \ldots \underline{\tilde{z}}_{N_R}^T]^T \qquad (QN_R) \times 1 \qquad (4.84)$$

and matrices $\underline{\underline{\Gamma}}$, $\underline{\underline{\check{\Lambda}}}_0$ and $\underline{\underline{\check{\Lambda}}}_{\tilde{h}}$ a are given by:

$$\underline{\underline{\check{\Lambda}}}_{\tilde{h}} := \begin{bmatrix} \underline{\underline{\Lambda}}_{\tilde{h}_{1,1}} & \cdots & \underline{\underline{\Lambda}}_{\tilde{h}_{1,N_T}} \\ \vdots & \ddots & \vdots \\ \underline{\underline{\Lambda}}_{\tilde{h}_{N_R,1}} & \cdots & \underline{\underline{\Lambda}}_{\tilde{h}_{N_R,N_T}} \end{bmatrix} \qquad (QN_R) \times (QN_T) \qquad (4.85)$$

$$\underline{\underline{\check{\Lambda}}}_0 := \underline{\underline{I}}_{N_R \times N_R} \otimes \underline{\underline{\Lambda}}_0 \qquad (QN_R) \times (QN_R) \qquad (4.86)$$

$$\underline{\underline{\Gamma}} := \underline{\underline{I}}_{N_R \times N_R} \otimes \underline{\underline{\gamma}} \qquad (QN_R) \times (QN_R) \qquad (4.87)$$

IMPACT OF THE NON-IDEAL FRONT-ENDS

The matrices $\underline{\underline{\check{\Lambda}}}_{\tilde{h}}$, $\underline{\underline{\check{\Lambda}}}_0$ and Γ have the following description and effect:

- $\underline{\underline{\check{\Lambda}}}_{\tilde{h}}$ (in Equation (4.85)) is a $(QN_R) \times (QN_T)$ block diagonal matrix made of diagonal blocks of size $Q \times Q$, each diagonal block $\underline{\underline{\Lambda}}_{\tilde{h}_{i,j}}$ containing the frequency response from antenna j to antenna i;

- $\underline{\underline{\check{\Lambda}}}_0$ (in Equation (4.86)) is a $(QN_R) \times (QN_R)$ block diagonal matrix, each block of size $Q \times Q$ being equal to $\underline{\underline{\Lambda}}_0$. The sub-blocks $\underline{\underline{\Lambda}}_0$ have both a phase rotation effect (4.21) and an amplitude effect (Equation (4.22));

- Γ (in Equation (4.87)) is a $(QN_R) \times (QN_R)$ block diagonal matrix, each block of size $Q \times Q$ being equal to $\underline{\underline{\gamma}}$. The sub-blocks $\underline{\underline{\gamma}}$ (Equation (4.23)) are responsible for ICI.

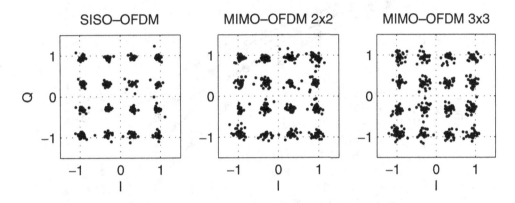

Figure 4.41 Constellation of SISO–OFDM, MIMO–OFDM 2×2 and MIMO–OFDM 3×3 with CFO and SCO, illustrating the higher degradation when more streams are transmitted. The MIMO–OFDM 2×2 and MIMO–OFDM 3×3 cases apply fully loaded SDM and use MMSE receivers.

The signal received on antenna j simplifies to:

$$\underline{\tilde{y}}_j = \underline{\underline{\gamma}} \cdot \underline{\underline{\Lambda}}_0 \cdot \sum_{i=1}^{N_T} [\underline{\underline{\Lambda}}_{\tilde{h}_{j,i}}] \cdot \underline{\tilde{x}}_i + \underline{\tilde{z}}_j \qquad (4.88)$$

Therefore, the effect of CFO and SCO on a MIMO–OFDM system will not only contain the effects described for a SISO–OFDM system (common phase rotation, phase slope across the sub-carriers and ICI originating from one symbol) but it will also contain an inter-stream interference because of the summation of N_T terms in Equation (4.88). This is illustrated in Figure 4.41 where we show the impact of the same amount of CFO and SCO on the constellation at the output of the equalizer, for three cases: SISO–OFDM (left plot), MIMO–OFDM 2×2 (center plot) and MIMO–OFDM 3×3 (right plot). The MIMO–OFDM 2×2 and MIMO–OFDM 3×3 cases apply fully loaded SDM and use MMSE receivers. Note that we have removed the common phase rotation here so that only ICI and amplitude effects are visible. The BER performance for the same SISO and MIMO set-up in high-delay spread channels is shown in Figure 4.42. The simulation was run for three values of CFO, namely 0, 3 and 10% of the sub-carrier spacing, corresponding to 0, 10 and 30 kHz, as in Section 4.1.2. Here again, the common phase rotation has been removed in the frequency domain, so that only the effect of ICI is visible. This kind of curve is very illustrative of system level trade-offs: at a CFO of approximately 3%, the compensation of the common phase rotation is sufficient, so that frequency domain tracking and compensation could be justified; for higher CFO, such as 10%, the frequency domain compensation is clearly inadequate and a time-domain compensation is needed.

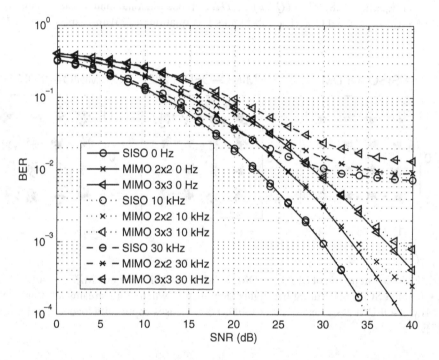

Figure 4.42 BER performance of SISO–OFDM, MIMO–OFDM 2×2 and MIMO–OFDM 3×3 with CFO and SCO. The SCO value corresponds to 40 ppm. Three different values of CFO are used, corresponding to 0, 3 and 10% of the sub-carrier spacing.

4.7.2 Sensitivity of STBC and MRC to CFO

STBC and MRC are routinely used in multiple-antenna set-ups to improve the link budget. They exploit the spatial dimension to provide higher diversity orders.[3] The diversity order of STBC with two transmit antennas (the 'Alamouti' scheme) is equal to $2 \times N_R$. The diversity

---

[3] The diversity order is equal to the slope of the BER curve at high SNR, in a log–log plot. This slope is actually the exponent affecting the SNR in the relation $BER = fn(itSNR)$.

IMPACT OF THE NON-IDEAL FRONT-ENDS

order of receive MRC is equal to N_R. It is interesting to compare the sensitivity of both techniques to analog non-idealities. We will consider STBC 2×1 and MRC 1×2, so that we compare two schemes with the same diversity order. In the ideal case, it is known that MRC has a gain of 3 dB over STBC, because STBC is not able to exploit the array gain at the transmit side, basically because the phase of the channel coefficients is not known at the point (the transmitter) where the spatial diversity is exploited.

In the MRC scheme, a single symbol at a time is transmitted. In the case of STBC, however, two symbols are transmitted simultaneously and are spatially separated at the receiver side (Section 2.3.1). The matched filter in Alamouti (1998) creates two perfectly orthogonal symbols so that each symbol can be detected independently of the other. When non-idealities are present, this exact orthogonality is broken. In the case of CFO, the STBC system model becomes for a single-carrier system, with two transmit and one receive antennas:

$$\begin{bmatrix} y[n] \\ y[n+1]^* \end{bmatrix} = \begin{bmatrix} \zeta & 0 \\ 0 & \xi \end{bmatrix} \cdot \begin{bmatrix} h_1 & h_2 \\ h_2^* & -h_1^* \end{bmatrix} \cdot \begin{bmatrix} d_1[i] \\ d_2[i] \end{bmatrix} + \begin{bmatrix} z[n] \\ z[n+1] \end{bmatrix} \quad (4.89)$$

where $i = \lfloor n/2 \rfloor$, $\zeta = e^{j\Delta\omega nQT}$ and $\xi = e^{-j\Delta\omega(n+1)QT}$ (T is the baseband sample rate, hence QT is the OFDM symbol duration). Note the '$-$' sign in the definition of ξ: the $y[n]$ and $y[n+1]$ are rotated in opposite directions with respect to each other. The signal at the output of the STBC matched filter then reads:

$$\begin{bmatrix} \hat{d}_1[i] \\ \hat{d}_2[i] \end{bmatrix} = \begin{bmatrix} |h_1|^2 + |h_2|^2 & 0 \\ 0 & |h_1|^2 + |h_2|^2 \end{bmatrix} \cdot \begin{bmatrix} d_1[i] \\ d_2[i] \end{bmatrix}$$

$$+ \begin{bmatrix} (\zeta-1)|h_1|^2 + (\xi-1)|h_2|^2 & h_1^* h_2 (\zeta - \xi) \\ h_1 h_2^* (\zeta - \xi) & (\zeta-1)|h_2|^2 + (\xi-1)|h_1|^2 \end{bmatrix} \cdot \begin{bmatrix} d_1[i] \\ d_2[i] \end{bmatrix}$$

$$+ \begin{bmatrix} w_1[i] \\ w_2[i] \end{bmatrix} \quad (4.90)$$

where w is a noise term. In Equation (4.90), we recognize the desired output and the noise; an additional term is responsible for a complex scaling of the desired output and the two terms '$h_1^* h_2 (\zeta - \xi)$' and '$h_1 h_2^* (\zeta - \xi)$' account for spatial interference between symbols $d_1[i]$ and $d_2[i]$. A visual interpretation of this phenomenon is proposed in Figure 4.43. The left plot shows the effect of CFO at the output of the STBC matched filter (without noise). The interference patterns are clearly visible. The MRC constellation (right plot) is only affected by a simple rotation since both receive antennas undergo the same rotation.

Very generally, in the presence of non-idealities, STBC will suffer from some degree of spatial interference. In MIMO systems affected by non-idealities, one can generally observe the following:

- MRC techniques are the least sensitive, STBC is more sensitive and SDM is the most sensitive;

- for SDM systems, fully loaded systems (number of streams equal to number of antennas) are more sensitive than underloaded systems.

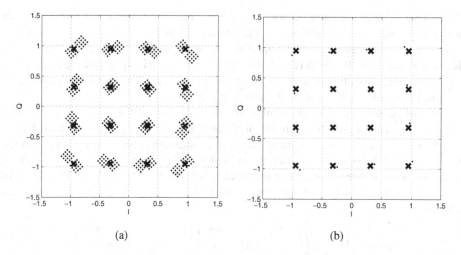

Figure 4.43 Impact of non-ideality on STBC 2 × 1 (a) and MRC 1 × 2 (b). The same CFO was applied on both systems. The higher degradation for STBC, originating from spatial interference, is visible.

4.7.3 Antenna mismatch and the reciprocity assumption

This sub-section analyzes specific transceiver requirements related to the reciprocity assumption for MIMO schemes with linear pre-coding. Systems belonging to the category of linearly pre-coded systems include Transmit MRC (MRC-TX), downlink SDMA (SDMA-DL) and joint transmit-receive (MIMO–TXRX). We will refer to the transmit directions as forward link and reverse link, the pre-coding processing being applied in the forward link. The reverse link will be used for channel estimation.

System model with the reciprocity assumption

We consider a wireless communication set-up with a forward link consisting of a transmitter (WT_A) with $N_T (N_T > 1)$ antennas and a receiver (WT_B) with $N_R (N_R \geq 1)$ antennas (Figure 4.44). WT_A intends to transmit $N_S (N_S \geq 1)$ streams applying a linear pre-coding scheme, thus requiring channel state information at the transmitter side (CSI-T). This CSI-T knowledge will be acquired through a reverse link transmission (often called *sounding*) with the assumption of channel reciprocity. This assumption is valid if two conditions are fulfilled (Kaiser *et al.* (2005), chapter 12): the duplex scheme is TDD and the downlink transmission occurs without significant delay after the reverse link channel estimation, compared to the coherence time of the channel. In what follows, we assume that the system is TDD and the channel is static, so that both conditions are met. The reverse link model for channel estimation is as follows (note that, in the reverse link, the transmitter has N_R antennas and the receiver has N_T antennas):

$$\underline{y} = \underline{\underline{H}}_{RL} \cdot \underline{d} + \underline{n} \quad (4.91)$$

IMPACT OF THE NON-IDEAL FRONT-ENDS

where:

- $\underline{d}$ is the $N_R \times 1$ vector of known transmitted symbols, for the purpose of reverse link channel estimation;
- $\underline{\underline{H}}_{RL}$ is the composite $N_T \times N_R$ reverse link channel matrix;
- $\underline{y}$ is the $N_T \times 1$ vector of received signals;
- $\underline{n}$ is the $N_T \times 1$ vector of noise samples.

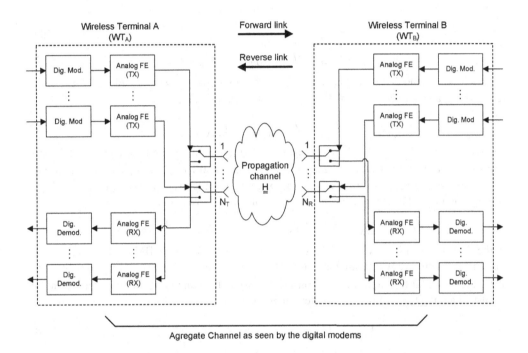

Figure 4.44 Wireless system block diagram showing the transmitter and receiver elements that are part of the channel. The forward and reverse link channels do not use the same transmitters and receivers.

Including the transmitters of WT_B and the receivers of WT_A, $\underline{\underline{H}}_{RL}$ can be expressed as:

$$\underline{\underline{H}}_{RL} = \underline{\underline{D}}_{TX,B} \cdot \underline{\underline{H}} \cdot \underline{\underline{D}}_{RX,A} \qquad (4.92)$$

where $\underline{\underline{D}}_{RX,A}$ and $\underline{\underline{D}}_{TX,B}$ are complex diagonal matrices containing, respectively, the N_T complex gains of the receivers in WT_A and the N_R complex gains of the transmitters in WT_B (throughout the analysis, we use the letter $\underline{\underline{D}}$ to emphasize that the matrices are diagonal). The matrices $\underline{\underline{D}}_{RX,A}$ and $\underline{\underline{D}}_{TX,B}$ are diagonal if we assume that there is no coupling between the different antenna branches. The matrix H contains the propagation channel itself, which is reciprocal. WT_A performs a channel estimation (based on the knowledge of the symbols transmitted in the reverse link), which provides the estimate $\underline{\hat{\underline{H}}}_{RL}$ (affected by $\underline{\underline{D}}_{TX,B}$ and $\underline{\underline{D}}_{RX,A}$). To simplify the analysis, we will assume that there are no channel estimation errors so that $\underline{\hat{\underline{H}}}_{RL} = \underline{\underline{H}}_{RL}$. In the forward link, the vector of the signals at the receive antennas is expressed as:

$$\underline{y} = \underline{\underline{H}}_{FL} \cdot \underline{\underline{F}}_{FL} \cdot \underline{x} + \underline{n} \qquad (4.93)$$

where $\underline{x}$ is the $N_S \times 1$ vector of transmitted symbols, $\underline{\underline{F}}_{FL}$ is the $N_T \times N_S$ pre-coding matrix (pre-filter), $\underline{\underline{H}}_{FL}$ is the composite $N_R \times N_T$ channel matrix, $\underline{n}$ is the $N_S \times 1$ vector of noise samples. An estimate of the transmitted symbols is obtained by means of a linear post-filtering:

$$\underline{\hat{x}} = \underline{\underline{G}}_{FL} \cdot \underline{y} \qquad (4.94)$$

$$= \underline{\underline{G}}_{FL} \cdot \underline{\underline{H}}_{FL} \cdot \underline{\underline{F}}_{FL} \cdot \underline{x} + \underline{\underline{G}}_{FL} \cdot \underline{n} \qquad (4.95)$$

in which $\underline{\underline{G}}_{FL}$ is the $N_S \times N_R$ receiver equalizer matrix (post-filter). Including the transmitters of WT_A and the receivers of WT_B, $\underline{\underline{H}}_{FL}$ can be expressed as:

$$\underline{\underline{H}}_{FL} = \underline{\underline{D}}_{TX,A} \cdot \underline{\underline{H}}^T \cdot \underline{\underline{D}}_{RX,B} \qquad (4.96)$$

where $\underline{\underline{D}}_{TX,A}$ and $\underline{\underline{D}}_{RX,B}$ are complex diagonal matrices containing, respectively, the complex gains of the transmitters in WT_A and of the receivers in WT_B. Equations (4.92) and (4.96) reveal that, because of the front-end complex gains, the reciprocity assumption is violated when some mismatch exist between antenna branches (Bourdoux *et al.* 2003):

$$\underline{\underline{H}}_{FL} = \underline{\underline{D}}_{TX,A} \cdot \underline{\underline{H}}^T \cdot \underline{\underline{D}}_{RX,B} \neq \underline{\underline{H}}_{RL}^T = \underline{\underline{D}}_{RX,A} \cdot \underline{\underline{H}}^T \cdot \underline{\underline{D}}_{TX,B} \qquad (4.97)$$

In the sequel, we will analyze the effect of non-reciprocity on some MIMO schemes. Without loss of generality, we will assume that the complex front-end gains $\underline{\underline{D}}_{RX,A}$ of the receivers in WT_A are all equal to 1 and that the non-reciprocity is introduced by complex values in the front-end gains $\underline{\underline{D}}_{TX,A}$ of the transmitters of WT_A as:

$$\underline{\underline{D}}_{TX,A} = (\underline{\underline{I}}_{N_T \times N_T} + \mathrm{diag}(\underline{\Delta\epsilon}) \cdot \mathrm{diag}(e^{j\underline{\Delta\phi}}) \qquad (4.98)$$

where the elements in vectors $\underline{\Delta\epsilon}$ and $\underline{\Delta\phi}$ are $\mathcal{N}(0, \sigma_{NR}^2)$. The parameter σ_{NR} is then a parameter that allows the amount of non-reciprocity to be controlled.

Post-processing SINR

Taking the system model of Equation (4.94) into account, it is possible to derive analytically the post processing SINR. We can rewrite Equation (4.94) as the simple product:

$$\underline{\hat{x}} = \underline{\underline{A}} \cdot \underline{x} + \underline{\underline{B}} \cdot \underline{n} \qquad (4.99)$$

where matrix $\underline{\underline{A}}$ captures the effect of the non-reciprocity matrix, the pre-filter, the propagation channel matrix and the post-filter, whereas matrix $\underline{\underline{B}}$ captures the post-filter only. The post processing SINR on stream j can be easily derived:

$$SINR_j = \frac{E_S \cdot \left[\mathrm{diag}(\underline{\underline{A}}) \cdot \mathrm{diag}(\underline{\underline{A}}^H)\right]_{jj}}{E_S \cdot \left[(\underline{\underline{A}} - \mathrm{diag}(\underline{\underline{A}})) \cdot (\underline{\underline{A}} - \mathrm{diag}(\underline{\underline{A}}))^H\right]_{jj} + \sigma_n^2 \left[\underline{\underline{B}} \cdot \underline{\underline{B}}^H\right]_{jj}} \qquad (4.100)$$

IMPACT OF THE NON-IDEAL FRONT-ENDS

where E_S is the average energy per symbol and σ_n^2 is the noise variance. In the numerator, we recognize the diagonal elements of $\underline{\underline{A}}$ that are the only ones to contribute to the desired symbol power; in the denominator, the first term captures all the inter-stream interference whereas the second term is due to the receiver noise affected by the post filter.

Impact of non-reciprocity on closed-loop SDM

In closed-loop SDM, the transmitter transmits several streams in parallel to a multi-antenna receiver. The transmitter uses the channel right singular vectors for the linear pre-coding in order to transmit the streams on the singular values of the channel. A power-loading strategy allows each stream to be weighted differently. At the receiver, the channel left singular vectors are used. This scheme is detailed in Section 2.3.2.

Non-reciprocity also impacts closed-loop SDM, but to a much lesser extent than downlink (DL-)SDMA. The immediate reason for this is that, although the linear-pre-coding performance is slightly degraded, resulting in a capacity loss, the multi-stream interference can still be completely eliminated by the spatial processing at the receiver. Indeed, the forward link channel estimation allows the receiver to estimate the channel, thus including the reciprocity error, and to compute a receiver post-filter adapted to the forward link actual channel. In other words, WT_A transmits with the linear pre-coder $\underline{\underline{D}}_{TX,A} \cdot \underline{\underline{F}}_{FL}$ but WT_B is able to estimate the aggregate channel $\underline{\underline{H}}_{FL} \cdot \underline{\underline{D}}_{TX,A} \cdot \underline{\underline{F}}_{FL}$ from which a better post-filter can be computed.

Figure 4.45 shows the BER of closed-loop SDM with non-reciprocity in frequency selective channels; in this case, the closed-loop SDM processing is applied per sub-carrier and perfect forward link channel estimation is assumed. The non-reciprocity was assumed to be constant in the frequency domain.

Impact of non-reciprocity on DL-SDMA

In DL-SDMA, the transmitter (in this case a base station) transmits several streams in parallel to several terminals. We will assume the terminals to have only one antenna, so each terminal receives only one stream. The transmitter applies a linear pre-coding in order to spatially separate the streams, so that each receiver receives only its own stream, free of interference from the other streams. The model in Equation (4.94) hence particularizes to $N_R = N_S$ and $\underline{\underline{G}}_{FL}$ being diagonal, containing the single antenna equalizer coefficients of each receiver. A possible strategy for the transmitter is the transmit MMSE pre-coding (TX-MMSE):

$$\underline{\underline{F}}_{FL,MMSE} = \frac{\underline{\underline{H}}_{FL}^H \cdot (\underline{\underline{H}}_{FL} \cdot \underline{\underline{H}}_{FL}^H + \sigma_n^2 \cdot \underline{\underline{I}}_{n_R \times n_R})^{-1}}{\|\underline{\underline{H}}_{FL}^H \cdot (\underline{\underline{H}}_{FL} \cdot \underline{\underline{H}}_{FL}^H + \sigma_n^2 \cdot \underline{\underline{I}}_{n_R \times n_R})^{-1}\|_{fro}} \quad (4.101)$$

Note that the denominator in Equation (4.101) is the Frobenius norm of the numerator and that it ensures that the MMSE pre-coder does not change the total transmit power:

$$\mathrm{Tr}\left[\underline{\underline{F}}_{FL,MMSE} \cdot \underline{\underline{F}}_{FL,MMSE}^H\right] = 1 \quad (4.102)$$

Note also that an estimate of the receiver noise variance σ_n^2 is needed for the MMSE pre-coder. If this noise variance is not known, it can either be assumed to be a reasonable

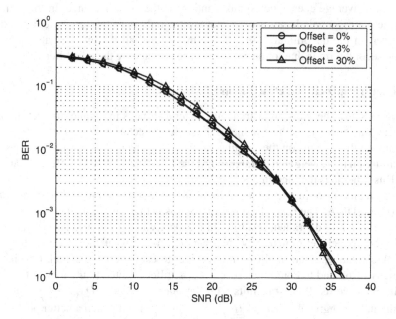

Figure 4.45 BER degradation of closed-loop SDM with non-reciprocity, in frequency selective channels. The amount of non-reciprocity varies from 0 to 30%.

working point (e.g. $\sigma_n^2 = 0.01$, which corresponds to SNR = 20 dB), or set to 0, which then corresponds to the zero-forcing solution.

Non-reciprocity has an significant impact on DL-SDMA systems. The immediate reason for this is that the streams are spatially separated at the transmitter side and that the receivers, being SISO receivers, have no mean to estimate the composite forward link channel including $\underline{\underline{D}}_{TX,A}$. Using Equation (4.100), we can assess the impact of non-reciprocity on the post processing SINR. Figure 4.46 shows the SINR degradation for various levels of non-reciprocity. To compute the SINR, we have assumed perfect downlink channel estimation: each single antenna receiver can then perfectly compensate the effect of non-reciprocity on its desired channel but is not able to compensate the inter-stream interference. Figure 4.47 shows the BER of DL-SDMA with non-reciprocity in frequency selective channels; in this case, the SDMA processing is applied per sub-carrier and perfect forward link channel estimation is assumed. The non-reciprocity was assumed to be constant in the frequency domain.

4.8 Multi-user systems

Like multi-antenna systems, the multi-user systems are multi-dimensional. Front-end non-ideality creates interference between the multiple dimensions (here the users).

The multi-user systems differ from the MIMO systems in that:

IMPACT OF THE NON-IDEAL FRONT-ENDS

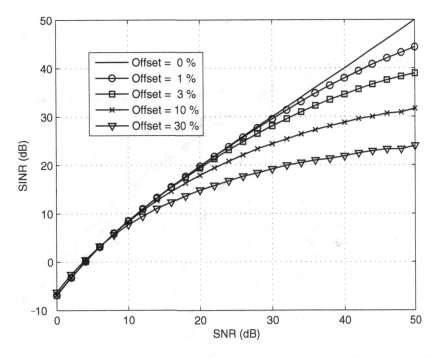

Figure 4.46 SINR degradation of DL-SDMA with non-reciprocity. The amount of non-reciprocity varies from 0 to 30%.

- the users are separated, based on a specific multi-access technology, such as CDMA or OFDMA. The impact of the front-end on the system performance depends on the chosen accessing scheme;

- because no hardware can be shared between the mobile terminals, the front-end non-ideality is always user-dependent which makes the estimation more complicated, since multiple parameters need to be estimated.

First, we compare the sensitivity of CDMA and OFDMA to the front-end non-ideality in the downlink of a cellular communication system. Second, we assess the impact of the user-dependent transmit front-end non-ideality and of the common receive front-end non-ideality on the performance of the uplink cellular communication system.

To perform the analysis, we focus on two non-idealities having a significantly different impact on the multi-carrier systems:

- the CFO, that generates interference between the neighboring carriers;

- the IQ imbalance, that generates interference of each carrier on its mirror carrier.

Figures 4.48 and 4.51 give the blocks diagrams of the downlink and uplink communication systems, respectively, considered in this section.

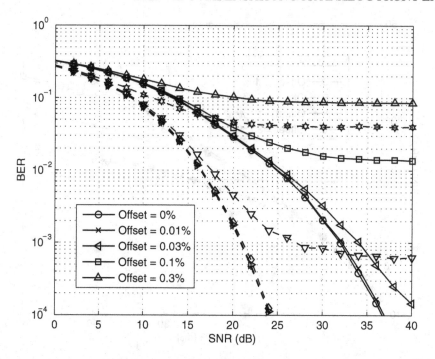

Figure 4.47 BER degradation of DL-SDMA with non-reciprocity, in frequency selective channels (solid lines: 2 × 2; dashed lines: 2 × 3). The amount of non-reciprocity varies from 0 to 30%.

4.8.1 MC-CDMA versus OFDMA

Figures 4.49 and 4.50 illustrate the performance of the MC-CDMA and OFDMA-based downlink communication systems in the presence of receive CFO and IQ imbalance, respectively. The system parameters are identical to those used in Section 2.5 to generate Figure 2.11. A fixed 1 kHz CFO and a fixed 2% phase mismatch have been considered in the simulations.

In the case of the MC-CDMA system, it can be observed that:

1. The user symbols are spread and interleaved on all carriers.

2. Therefore, any front-end non-ideality creates ISI and MUI and the BER performance is always limited by the interference.

3. Since the user signals are recovered by de-spreading with the CDMA codes, the impact of the non-ideality is partially averaged out.

In the case of the OFDMA system with a distributed carrier allocation, it can be observed that:

1. The user symbols are placed on a set of equally spaced carriers.

IMPACT OF THE NON-IDEAL FRONT-ENDS

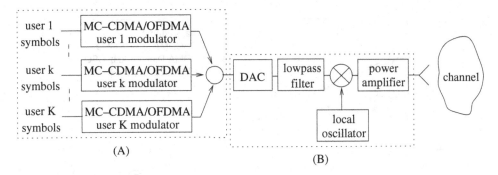

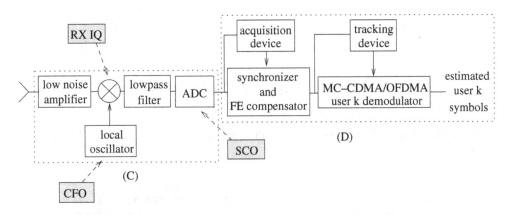

Figure 4.48 Downlink system block diagram: (A) transmit digital transceiver; (B) transmit analog front-end; (C) receive analog front-end; (D) receive digital transceiver.

2. Therefore, IQ imbalance potentially creates a significant amount of ISI and MUI, limiting the BER performance (note, however, that the single-user system is quite robust to IQ imbalance because the user symbols have been placed by chance on a set of non-mirror carriers).

3. When the number of users is small (one user in the example), the system is particularly robust against the CFO because guard carriers protect the symbols from significant interference. This nice property is obviously lost when the number of users is increasing (10 users in the example).

In the case of the OFDMA system with a localized carrier allocation, it can be observed that:

1. The user signals are placed on a set of adjacent carriers.

2. Therefore, the CFO generates always a high level of ISI, limiting the BER performance. At high SNR, when the interference dominates over the noise, the OFDMA localized system performs worse than the MC-CDMA system.

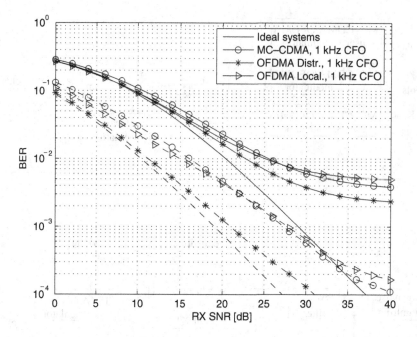

Figure 4.49 Impact of the CFO on the downlink performance (dashed curves: 1 user; solid curves: 10 users).

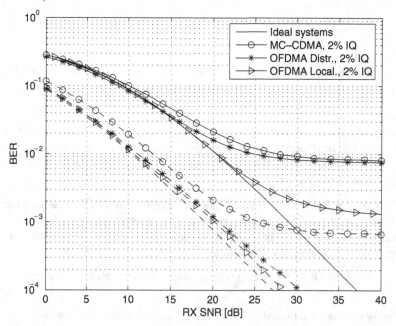

Figure 4.50 Impact of the IQ imbalance on the downlink performance (dashed curves: 1 user; solid curves: 10 users).

IMPACT OF THE NON-IDEAL FRONT-ENDS

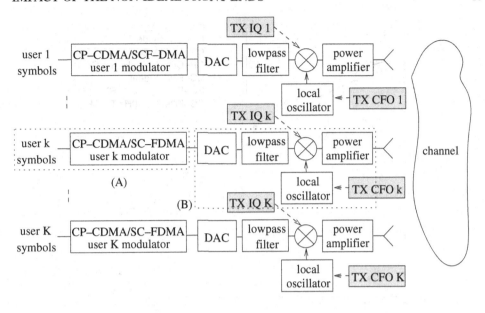

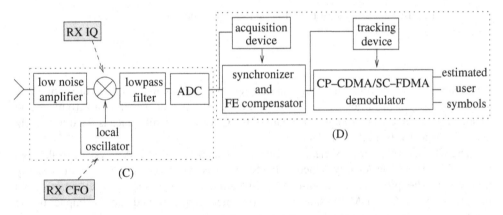

Figure 4.51 Uplink system block diagram: (A) transmit digital transceiver; (B) transmit analog front-end; (C) receive analog front-end; (D) receive digital transceiver.

3. When the number of users is small (one user in the example), the system is particularly robust against the IQ imbalance because the user symbols only interfere on unused carriers. This nice property is obviously also lost when the number of users is increasing.

4.8.2 User dependent non-idealities

Figures 4.52 and 4.53 illustrate the performance of the CP-CDMA and distributed SC-FDMA-based uplink communication systems in the presence of transmit/receive CFO and

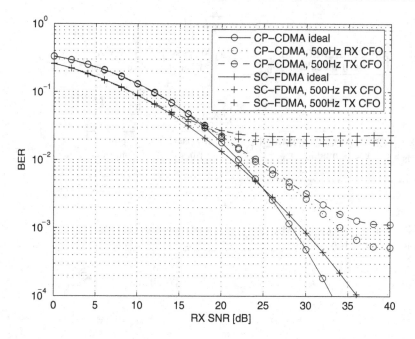

Figure 4.52 Impact of the CFO on the uplink performance (10 users).

IQ imbalance, respectively. The system parameters are identical to those used in Section 2.5. The front-end non-idealities are assumed to follow a Gaussian distribution of zero mean and of specified variance. The root mean square of the CFO and IQ imbalance is fixed to 500 Hz and 1%, respectively, in the simulations.

The CP-CDMA system is clearly more robust than the SC-FDMA system to the two types of front-end non-ideality. Whereas the SC-FDMA system loses its good orthogonality property in the presence of front-end non-ideality, therefore strongly deteriorating its performance, the CP-CDMA system is *a priori* interference-limited and the impact of the front-end non-ideality on its performance is less pronounced.

On the other hand, transmit CFO has a stronger impact than receive CFO on the system performance. Since the average performance in a multi-dimensional system is always limited by the worst dimension, the chance of having a high CFO value among the different users in the uplink is higher than that in the downlink.

Conversely, receive IQ imbalance has a stronger impact than transmit IQ imbalance on the system performance. In both cases, the symbol received on one carrier suffers from interference coming from the symbol located on the mirror carrier. When the IQ imbalance is generated at the receiver, the interfering symbol is multiplied with the channel coefficient of the mirror carrier. Therefore, a high-gain interfering carrier can destroy the symbol estimation on a low-gain carrier. When the IQ imbalance is generated at the transmitter, the interfering symbol is multiplied with the channel coefficient of the carrier itself (like the symbol of interest). Therefore, all carriers suffer from interference proportional to the carrier gain itself.

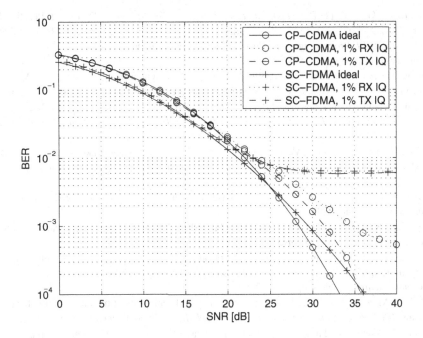

Figure 4.53 Impact of the IQ imbalance on the uplink performance (10 users).

References

Alamouti, S. (1998) A simple transmit diversity technique for wireless communications. *IEEE Journal on Selected Areas in Communications* **16**(8), 1451–1458.

Armada, A. and Calvo, M. (1998) Phase noise and sub-carrier spacing effects on the performance of an OFDM communication system. *IEEE Communication Letters* **2**(1), 11–13.

Bourdoux, A., Come, B. and Khaled, N. (2003) Non-reciprocal transceivers in OFDM/SDMA systems: impact and mitigation. *Proceedings of IEEE Radio and Wireless Conference*, pp. 183–186.

Horlin, F., De Rore, S., Lopez-Estraviz, E., Naessens, F. and Van der Perre, L. (2006) Impact of frequency offsets and IQ imbalance on MC-CDMA reception based on channel tracking. *IEEE Journal on Selected Areas in Communications* **24**(6), 1179–1188.

Jang, J. and Lee, K. (1999) Effects of frequency offset on MC/CDMA system performance. *IEEE Communications Letters* **3**(7), 196–198.

Kaiser, T., Bourdoux, A., Boche, H., Fonollosa, J., Andersen, J. and Utschick, W. (2005) *Smart Antennas State of the Art*. Eurasip.

Kim, Y., Bang, K., Choi, S., You, C. and Hong, D. (1999) Effect of carrier frequency offset on performance of MC-CDMA systems. *IEEE Electronic Letters* **35**(5), 378–379.

Moeneclaey, M. (1997) The effects of synchronization errors on the performance of orthogonal frequency-division multiplexed (OFDM) systems. *IEEE Proceedings of the COST 254 on Emergent Techniques for Communication Terminals*.

Ochiai, H. and Imai, H. (2000) On clipping for peak power reduction of OFDM signals. *Proceedings of IEEE GLOBECOM*, pp. 731–735.

Pollet, T., Bladel, M.V. and Moeneclaey, M. (1995) BER sensitivity of OFDM systems to carrier frequency offset and weiner phase noise. *IEEE Transactions on Communications* **43**(1), 191–193.

Rabiner, L. and Gold, B. (1975). *Theory and Application of Digital Signal Processing*. Prentice-Hall.

Reinhardt, S. and Weigel, R. (2004) Pilot aided timing synchronization for SC-FDE and OFDM: A comparison. *IEEE Proceedings of the International Symposium on Communications and Information Technology*, vol. 1, pp. 628–633.

Steendam, H. and Moeneclaey, M. (1999a) The effect of synchronization errors on MC-CDMA performance. *IEEE Proceedings of International Conference on Communications*, pp. 1510–1514.

Steendam, H. and Moeneclaey, M. (1999b) The sensitivity of a flexible form of MC-CDMA to synchronization errors. *IEEE Proceedings of Vehicular Technology Conference*, vol. 4, pp. 2208–2212.

Steendam, H. and Moeneclaey, M. (2002) The effect of clock frequency offset on downlink MC-DS-CDMA. *IEEE Proceedings of 7th Symposium on Spread Spectrum Technology and Applications*, vol. 1, pp. 113–117.

Tomba, L. (1998) On the effect of wiener phase noise in OFDM systems. *IEEE Transactions on Communications* **46**(5), 580–583.

Tomba, L. and Krzymien, W. (1999) Sensitivity of the MC-CDMA access scheme to carrier phase noise and frequency offset. *IEEE Transactions on Vehicular Technology* **48**(5), 1657–1665.

Tubbax, J., Come, B., Van der Perre, L., Deneire, L., Donnay, S. and Engels, M. (2003) Compensation of IQ imbalance in OFDM systems. *IEEE Proceedings of International Conference on Communications*, vol. 5, pp. 3403–3407.

Yu, H., Kim, M. and Ahm, J. (2003) Carrier frequency and timing offset tracking scheme for SC-FDE systems. *IEEE Proceedings of Personal, Indoor and Mobile Radio Communications*, vol. 1, pp. 1–5.

Zamorano, J., Nsenga, J., Thillo, W.V., Bourdoux, A. and Horlin, F. (2007) Impact of phase noise on OFDM and SC-CP. *IEEE Proceedings of Globecom*.

5

Generic OFDM System

Since the end of the 1990s, OFDM and OFDMA have become two of the most widely used modulation techniques. They can be found now in major digital television and radio broadcast standards (DVB-T/H, DAB, ISDB-T, DMB), in major outdoor cellular systems (IEEE802.16-2004, IEEE802.16e-2005, 3GPP-LTE), in WLANs (IEEE802.11a/g/n) and WPANs (IEEE802.15.3c, Ecma 368). The reasons for the success of OFDM are many: elegant multi-path mitigation in the frequency domain, very flexible multiplexing and multiple access by sub-carrier allocation, simple per-sub-carrier processing in MIMO-OFDM, etc.

OFDM has been described in detail in Chapter 2 and its sensitivity to non-idealities was addressed in Chapter 4. The goal of this chapter is to describe how non-idealities can be tackled in OFDM transmitters or receivers. We will start by describing a generic OFDM system, with a frame format containing a preamble and pilots, as with most modern burst transmission systems. This generic OFDM system will then be used to show how the OFDM modem receiver is designed to process the OFDM burst and, at the same time, compensate for the non-idealities introduced by the front-end. We will delve into the details of the acquisition schemes (burst detection, timing, CFO estimation), estimation of the channel and front-end parameters and tracking of the CFO, SCO and phase noise.

5.1 Definition of the generic OFDM system

For the purpose of describing how estimation and compensation of front-end non-idealities can be performed, we introduce a generic OFDM system. For convenience, but without loss of generality, this system is inspired from the OFDM PHY layer of the IEEE802.11a/g system with some simplifications.

We consider a single link symmetrical system, without distinction between forward or reverse direction, in a typical multi-path indoor environment. The multi-path propagation channel is assumed to be frequency selective and static. (NB: time-varying effects will be considered in this chapter but are not due to propagation effects.) Another assumption that we will make throughout is that the excess delay of the channel is smaller than the length of the cyclic prefix, so that no inter-block interference (IBI) occurs.

5.1.1 Frame description

The transmission format of the physical layer is a frame, which consists of a preamble part and a data or payload part. The frame structure is illustrated in Figure 5.1.

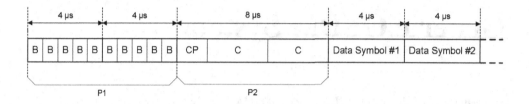

Figure 5.1 OFDM preamble structure.

Preamble description

The preamble is 16 μs long. It consists of two equal length parts. The first part of the preamble (P1), which is 8 μs long, consists of 10 identical short symbols (called 'B', 800 ns each) and is designed to enable burst detection and coarse acquisition of timing and carrier frequency offset. The second part of the preamble (P2), which is also 8 μs long, consists of two identical long symbols (called 'C', 3.2 μs each) and is designed to enable fine acquisition of timing and CFO as well as channel estimation. In P2, a double length CP of 1.6 μs is added so that P2 is completely phase continuous and cyclic over its full length. A greater robustness to timing inaccuracy after coarse timing acquisition results. In the following, we will use the physical index to represent the sub-carrier index. The 64 sub-carriers have an integer index in the interval $[-31, 32]$ with a frequency index 0 corresponding to DC. The detail of the sub-carrier use in the preamble and payload part of the frame is illustrated in Figure 5.2. Because of the repetition in the time domain (4 × 800 ns in one nominal 3.2 μs symbol duration), P1 is fully described by 12 frequency domain symbols with a sub-carrier index that are multiples of four. Note that their amplitude is $\sqrt{52/12} = \sqrt{13/3}$ higher than that of the sub-carrier in P2 or in the payload, so that the power in one symbol of P1 is the same as that in P2 and in the payload symbols. Finally, the sub-carriers with an index above 26 or under -26 cater for guard bands in the PSD of the transmitted signal, which prevents adjacent channel interference and eases filtering in the TX and RX parts.

So far, we have only discussed the time and frequency localization properties of parts P1 and P2 of the preamble. The (complex) values carried by each sub-carrier in the preamble can be freely chosen. They are usually selected to result in a low PAPR during the preamble part in order to avoid nonlinearity degradation during synchronization and channel estimation. The instantaneous absolute value of the preamble is shown in Figure 5.3. The PAPR of P1 is 2.09 dB and the PAPR of P2 is 3.17 dB.

GENERIC OFDM SYSTEM

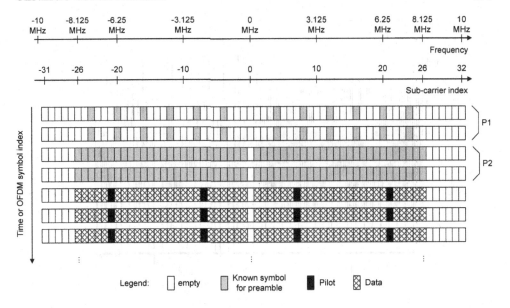

Figure 5.2 Detail of the sub-carrier allocation in the OFDM frame.

Table 5.1 Generic OFDM system parameters.

Carrier frequency	5 GHz
Signaling rate	20 MHz
Number of sub-carriers	$Q = 64$
sub-carrier spacing	312.5 kHz
Occupied carriers	48 (data) +4 (pilot)
Symbol length	$Q = 64$ samples or 3.2 µs
CP length	$L = 16$ samples or 0.8 µs
Total symbol length	$(K = Q + L) = 80$ samples or 4 µs
Constellation	QPSK, 16QAM, 64QAM

Data part description

The OFDM modulation is similar to that already introduced in Section 2.1. The parameters of our generic OFDM PHY are detailed in Table 5.1. A total of 52 sub-carriers are used per channel, where 48 sub-carriers carry actual data and four sub-carriers are pilots, which facilitate CFO and SCO tracking and phase noise compensation for coherent demodulation. The duration of the guard interval is equal to 800 ns, which is sufficient to enable good performance on most indoor multi-path wireless channels. Note that our analysis will be made for uncoded systems; hence, the coding scheme and coding rates are not necessary for our discussion.

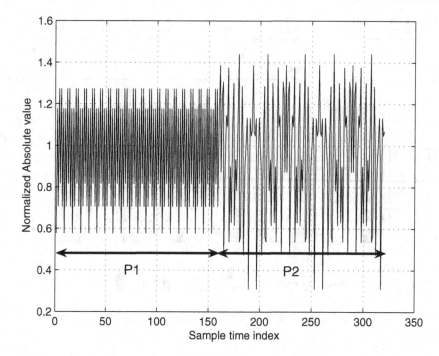

Figure 5.3 Preamble instantaneous absolute value. The repetition of 10 'B' sequences can be recognized on the left and the two receptions of the 'C' sequence, with the long CP, is visible on the right. Note the low PAPR during the preamble.

5.1.2 OFDM receiver description

The frame described in the previous section was designed to make acquisition and demodulation possible. The following processes must be successfully implemented.

1. Burst detection: the receiver must distinguish between receiver noise and the arrival of a desired burst. Other signals should also not trigger the receiver.

2. AGC setting: once the receiver has decided that a desired signal is being received, its amplitude must be measured in order to set the (analog) gain of the receiver so that the signal lies ideally in the dynamic range of the ADC.

3. Timing acquisition: the symbol timing must be estimated so that the OFDM blocks are correctly extracted from the burst. This is quite critical, since any timing error has a penalty on the system performance: too early timing consumes part of the cyclic prefix and reduces resistance to multi-path; too late timing causes inter-symbol interference that cannot be compensated.

4. CFO acquisition: the CFO must be estimated so that it can be compensated for. The residual CFO after the acquisition and compensation must be small enough to be tracked and compensated for in the tracking loop.

GENERIC OFDM SYSTEM

5. Channel estimation: the multi-path channel coefficients must be estimated.

6. Demodulation and equalization: the payload data must be extracted, equalized and demodulated.

7. Tracking: residual synchronization errors at the end of the preamble must be tracked and compensated during the payload data processing.

The structure of a generic OFDM receiver, designed to perform these operations on the received OFDM frame signals as described in Section 5.1.1 is illustrated in Figure 5.4. Its operation is described in the next two sections.

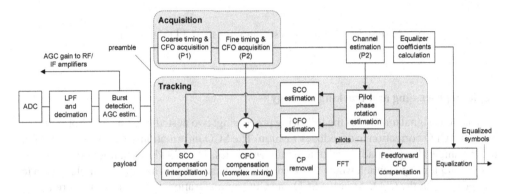

Figure 5.4 Generic OFDM receiver block diagram, showing the preamble path (upper branch) and the payload path (lower branch).

Acquisition strategy

Figure 5.5 shows a typical partitioning of the preamble for the different processes described in the previous section. Note that the time devoted to the burst detection and AGC setting must be traded off against the time for coarse acquisition. The oversampled signal from the ADC is first lowpass filtered and decimated to bring the signal at the nominal OFDM sampling rate (20 Msps). Of course, because of the SCO offset, this sampling rate is not exactly equal to the transmitter sampling rate. Next, the presence of the OFDM signal must be detected and, once a signal is detected, the gain of the receiving chain must be adjusted so that the signal falls ideally in the ADC dynamic range. Note that the gain setting must happen *before* the acquisition and channel estimation takes place. A few 'B' sequences of P1 are consumed for burst detection and AGC setting, typically three 'B' sequences. The remaining part of P1 is used for CFO and timing acquisition. The repetitive nature of P1 is used for these purposes. The coarse timing and CFO acquisition on P1 are used to compensate P2. P2 is then used to refine the timing and CFO acquisition. Next, P2 is corrected with the estimated CFO and channel estimation takes place. The channel estimation can then be used to compute the equalizer coefficients, which remain constant for the rest of the burst in static channels. Note that, at this point in time, the initial SCO offset in the preamble is compensated for by the channel estimation. The SCO – and any residual CFO and phase noise – will have to be taken care of by a tracking loop that runs on the payload part of the frame.

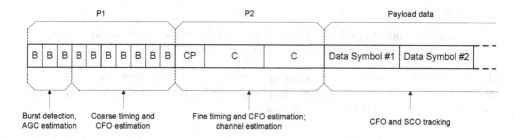

Figure 5.5 Acquisition strategy for the OFDM system.

Payload processing and tracking strategy

The payload signal undergoes the following processing: two nested tracking loops provide the SCO and CFO correction in a feedback fashion. The SCO information is fed to an interpolator that resamples the payload signal at non integer indices. The CFO information is transformed into a complex exponential, which is mixed element-wise with the incoming sample stream to compensate for the CFO. After the SCO and CFO compensation, the cyclic prefix is removed from the block and the FFT is computed. The FFT output provides the transmitted frequency domain symbols, affected by the channel frequency response and by phase noise and any residual CFO and/or SCO. To estimate and compensate for this, the pilots are extracted here and their phase is compared with that of the ideal pilots affected by the channel frequency response. The mean of the frequency domain phase differences is used to estimate the CFO whereas the slope of the frequency domain phase differences is used to estimate the SCO. The estimated SCO is fed to the SCO compensation block, which interpolates the subsequent samples. The estimated CFO determines the frequency of a complex exponential, which is mixed with the interpolated signal. Note that these two compensations affect the subsequent OFDM symbols.[1] Hence, it is also necessary to compensate the current OFDM symbol in a feedforward fashion. This is, for practical values of CFO and SCO, only needed for the CFO. The last step is the equalization.

5.2 Burst detection

The burst detection problem is a simple binary hypothesis testing problem that can be expressed as follows: calling $y[n]$, $x[n]$ and $w[n]$ the samples of the received signal, the desired signal to be detected and the noise, respectively; the receiver has to decide between hypotheses H_0 and H_1. When only noise is present:

$$H_0 : y[n] = w[n] \tag{5.1}$$

---

[1] It is also possible to compensate for CFO and SCO and then recompute the FFT. This iterative CFO/SCO compensation has better performance but increases complexity, power consumption and latency.

GENERIC OFDM SYSTEM

whereas when both noise and the desired signal are present:

$$H_1 : y[n] = x[n] + w[n] \tag{5.2}$$

The performance of this process can be characterized by the probability of false alarm P_{FA}, which is the probability of (erroneously) deciding H_1 when H_0 is true and the probability of detection, which is the probability of (correctly) deciding H_1 when H_1 is true. For this analysis, we define the variance of the transmitted symbols as σ_x^2 and the noise variance as σ_w^2.

5.2.1 Energy-based detection

The simplest method to detect the presence of a signal is by energy detection. The receiver accumulates the squared amplitudes of N consecutive incoming complex samples, producing a measure $z = \sum_{k=1}^{N} |y[k]|^2$ of the incoming energy. Under H_0 and if the noise is Gaussian distributed, the distribution of the energy detector output z is chi-squared with 2N degrees of freedom. Its PDF can be expressed as:

$$P_Z(z) = \frac{1}{\sigma_w^{2N} 2^N (n!)} z^{N-1} e^{-z/2\sigma_w^2} \tag{5.3}$$

For a given detection threshold γ, the false alarm probability is equal to the complementary cumulative distribution function of Z under the noise-only hypothesis (H_0)

$$P_{FA}(\gamma) = \int_{\gamma}^{\infty} P_{Z|H_0}(z) \, dz = e^{-\gamma/2\sigma_w^2} \sum_{k=0}^{N-1} \frac{1}{k!} \left(\frac{\gamma}{2\sigma_w^2}\right)^k \tag{5.4}$$

Equation (5.4) can be used to derive the threshold γ, for a given P_{FA}. The P_{FA} is plotted in Figure 5.6 (top) for 16, 32 and 48 samples, respectively, which corresponds to 1, 2 or 3 'B' sequences of P1. These curves were drawn with the assumption of the samples of $w[n]$ being $\mathcal{CN}(0, 1)$. Assuming that the signal samples are also Gaussian distributed (which is approximately true for OFDM signals by virtue of the central limit theorem), the detection probability P_D is also derived from the complementary cumulative distribution function of Z:

$$P_D(\gamma, SNR) = \int_{\gamma}^{\infty} P_{Z|H_1}(z) \, dz = e^{-z/2\sigma_1^2} \sum_{k=0}^{N-1} \frac{1}{k!} \left(\frac{z}{2\sigma_1^2}\right)^k \tag{5.5}$$

Note that in Equation (5.5) the variance σ_1^2 is the sum of the signal and noise variances, whereas the variance in Equation (5.4) is just the noise variance. This is the reason why P_D also depends on the SNR. The P_D is plotted in Figure 5.6 (three lower plots) for the case of 16, 32 and 48 integrated samples.

As an example, consider the case of 48 samples and a P_{FA} of 10^{-7}; a threshold value of about 95 is needed. Figure 5.6 (bottom) shows the detection probability under H_1 and for 48 accumulated samples. With the threshold value set to about 95 to guarantee the P_{FA} of 10^{-7}, the detection probability P_D is close to 1 at 2 dB SNR and 0.6 at 0 dB SNR. In practice, an even higher threshold would be more practical to avoid false alarms; probably around 100, which still gives reasonable performances around 0 dB SNR. We also note from Figure 5.6

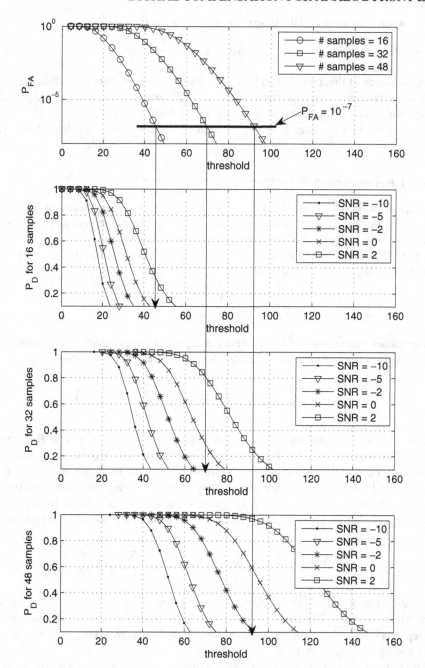

Figure 5.6 P_{FA} and P_D for the energy detection method in burst detection. The cases of 16, 32 and 48 samples are considered for the P_{FA} (top figure). The required P_{FA} determines the threshold. The P_D corresponding to a P_{FA} of 10^{-7} is shown in the three lower figures, for 16, 32 and 48 samples and with the SNR as a parameter.

GENERIC OFDM SYSTEM

that 16 or 32 accumulated samples do not provide a sufficient P_D around 0 dB SNR. If good detection performance is desired with 16 or 32 samples, either a higher SNR is needed or the P_{FA} requirements must be relaxed. A possible drawback of the energy detection method is that it will trigger the receiver for *any* energy entering the receiver, whereas only the desired preamble should do this. This is not a desired behavior when other wireless systems are being used, for example in the industrial scientific and medical (ISM) bands. Hence, a method based on the specific structure of the preamble, namely the repetitions, can be exploited.

5.2.2 Auto-correlation-based detection

An improved detection method can be conceived by exploiting the repetition in the preamble. Instead of summing the energy ($z = \sum_{k=1}^{N} |y[k]|^2$), the idea is to sum the product of samples separated by the repetition interval L of the B sequences: $z = \sum_{k=1}^{N} y[k] \cdot y^*[k+L]$. This is clearly an estimate of the auto-correlation for a delay L. We must have a closer look at the resulting product in order to derive its statistics. Defining

$$x[k] := x_r[k] + jx_i[k] \tag{5.6}$$

$$w[k] := w_r[k] + jw_i[k] \tag{5.7}$$

a single product reads:

$$\begin{aligned}
& y[k] \cdot y^*[k+L] \\
&= x_r[k]x_r[k+L] + x_i[k]x_i[k+L] + x_r[k]w_r[k+L] + x_i[k]w_i[k+L] \\
&\quad + w_r[k]x_r[k+L] + w_i[k]x_i[k+L] + w_r[k]w_r[k+L] + w_i[k]w_i[k+L] \\
&\quad + j \cdot (-x_r[k]w_i[k+L] + x_i[k]w_r[k+L] - w_r[k]x_i[k+L] + w_i[k]x_r[k+L] \\
&\quad - w_r[k]w_i[k+L] + w_i[k]w_r[k+L])
\end{aligned} \tag{5.8}$$

In Equation (5.8), we recognize the following contributions:

- a deterministic component due to the product of $x[k]$ with $x[k+L]$. Globally, this will produce a real constant contribution equal to $N\sigma_x^2$;

- the cross-product of $x[k]$ with $w[k+l]$ or $x[k+L]$ with $w[k]$. Note that X and W are independently distributed. Assuming that the $x[k]$ are Gaussian distributed, the resulting product has a distribution given by:

$$P_{XW}(z) = \frac{K_0(\frac{|z|}{\sigma_x \sigma_w})}{\pi \sigma_x \sigma_w} \tag{5.9}$$

where $K_0(z)$ is the modified Bessel function of the second kind with parameter 0. Since X and W have zero mean and are symmetrically distributed, their product has also zero mean. The variance of their product is given by $\int_{-\infty}^{\infty} P_{XW}(z)\, dz$ and is equal to $\sigma_x^2 \sigma_w^2$;

- the terms due to the product of $w[k]$ with $x[k+L]$. These are also products of independently distributed Gaussian variables. Their mean is zero and their variance is equal to σ_w^4.

The term derived in Equation (5.8) is summed N times (e.g. 16, 32 or 48 times) so that we can apply the central limit theorem. Hence, and by careful examination of Equation (5.8), the real and imaginary part of z have the following distributions:

- real part of z under H_0: $\Re(z) \sim \mathcal{N}(0, 2N\sigma_w^4)$;
- real part of z under H_1: $\Re(z) \sim \mathcal{N}(2N\sigma_x^2, 4N\sigma_x^2\sigma_w^2 + 2N\sigma_w^4)$;
- imaginary of z part under H_0: $\Im(z) \sim \mathcal{N}(0, 2N\sigma_w^4)$;
- imaginary of z part under H_1: $\Im(z) \sim \mathcal{N}(0, 4N\sigma_x^2\sigma_w^2 + 2N\sigma_w^4)$.

Since the useful part of x is solely present in the real part of Z, the detector is simply $z = \Re(\sum_{k=1}^{N} y[k] \cdot y^*[k+L])$. Given the mean and variance of Z under H_0 and H_1, its P_{FA} and P_D are given by the complementary cumulative distribution function of a Gaussian:

$$P_{FA}(\gamma) = \int_{\gamma}^{\infty} P_{Z|H_0}(z)\,dz = \frac{1}{2}\mathrm{erfc}\left(\frac{\gamma}{\sqrt{2}\sigma_{Z0}}\right) \quad (5.10)$$

$$P_D(\gamma, SNR) = \int_{\gamma}^{\infty} P_{Z|H_1}(z)\,dz = \frac{1}{2}\mathrm{erfc}\left(\frac{\gamma - m_{Z1}}{\sqrt{2}\sigma_{Z1}}\right) \quad (5.11)$$

In Equations (5.10) and (5.11), we have defined the mean of Z under H_1 as m_{Z1}, the noise variance under H_0 as σ_{Z0} and the noise variance under H_1 as σ_{Z1}. The P_{FA} and P_D of Z with the auto-correlation estimator is plotted in Figure 5.7. By comparison with Figure 5.6, we observe the improved performance of this detection method: it achieves similar performances with 32 samples as the energy detector with 48 samples. The superior performance in rejecting other wireless systems (avoiding false alarms) is illustrated in Figure 5.8. Because the auto-correlation detector is designed to detect signals with a repetition of 16 samples, its output returns to a low value at the end of the desired part of the preamble (P1). The energy detector, however, is not able to discriminate the signals. Hence, it keeps detecting energy during the remaining part of the preamble (P2) and during the data portion of the frame.

5.3 AGC setting (amplitude estimation)

The received signal consists of noise-only samples before the burst arrives and of the 'B' sequences of P1 convolved with the channel impulse response at the beginning of the burst. Because the channel response is by definition unknown during this phase, the best and simple estimator is the RMS amplitude estimator that estimates A_{RMS} by means of N input samples of the received signal $y(t)$ as

$$A_{RMS} = \sqrt{\frac{1}{N}\sum_{k=1}^{N} |y[k]|^2} \quad (5.12)$$

Figure 5.9 shows the PDF of the squared RMS amplitude estimation for various levels of SNR. In this simulation, one 'B' sequence was convolved with a high-delay spread channel and corrupted by AWGN noise. The noise level with variance 1 was held constant and the signal power at the RX antenna was changed to fix the desired SNR. Then, 16 samples were

GENERIC OFDM SYSTEM

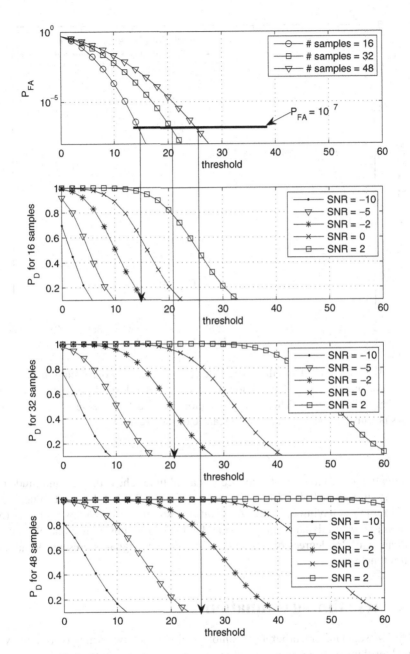

Figure 5.7 P_{FA} and P_D for the auto-correlation method in burst detection. The cases of 16, 32 and 48 samples are considered for the P_{FA} (top figure). The required P_{FA} determines the threshold. The P_D corresponding to a P_{FA} of 10^{-7} is shown in the three lower figures, for 16, 32 and 48 samples and with the SNR as a parameter.

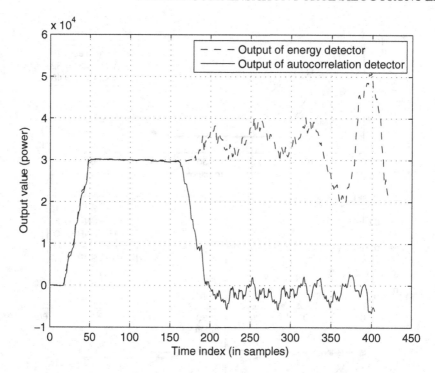

Figure 5.8 Detector output for the energy detector and the auto-correlation detector. The auto-correlator output returns to very low values when the desired preamble is not present (around sample index 160). The energy detector keeps a high value when the preamble disappears because it does not exploit the feature of the preamble.

used in Equation (5.12) to estimate the received amplitude. The estimated amplitude is quite accurate since more than 99% of the estimates are within ±3 dB of the exact value, justifying the rule-of-thumb of having one extra ADC bit of resolution for AGC inaccuracies (one bit corresponds to 6 dB of dynamic range). Note that the histograms correspond closely to Chi-square PDFs with 2N degrees of freedom, even when the signal is present, because of the central limit theorem.

5.4 Coarse timing estimation

Many OFDM standards incorporate preambles suitable for two kinds of timing synchronization algorithms: auto-correlation and cross-correlation. The auto-correlation algorithms have been studied extensively by Moose (1994) and Schmidl and Cox (1997). This class of algorithms performs an auto-correlation (AC) of the received signal to detect a repeated pattern. The cross-correlation (XC) algorithm has been proposed in Hazy and El-Tanany (1997) and Tufvesson et al. (1999). This class of algorithms performs a XC of the received signal with a known training symbol. The AC has a better robustness against multi-path since

GENERIC OFDM SYSTEM

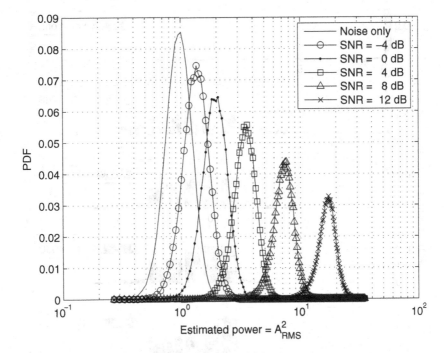

Figure 5.9 Histogram of estimated power for various SNR levels. A power of 1 corresponds to noise only.

the 'reference' in the AC is the signal itself, affected by multi-path. However, the XC uses a *known* reference and is therefore more robust in situations of low SNR. Both methods have interesting by-products: the AC easily provides an estimate of the CFO while the XC output provides the channel impulse response.

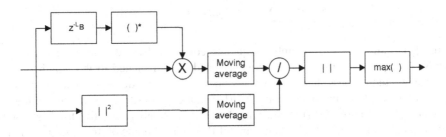

Figure 5.10 Auto-correlation timing structure. The bottom path is a normalization.

The block diagram of an AC algorithm is shown in Figure 5.10. Note that a power normalization (the bottom path) is added as this makes the packet detection more robust

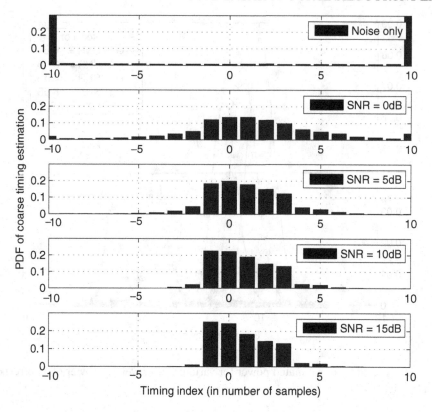

Figure 5.11 Histograms showing the performance of the timing estimation based on the auto-correlation of the received P1.

to errors in the initial AGC stage. The AC algorithm is based on the following function:

$$AC[n] = \sum_{m=0}^{M-1} r[n+m]r^*[n+m-L_B] \qquad (5.13)$$

where $r[n]$ is the received preamble sequence corrupted by the multi-path channel and AWGN, L_B is the length of the B sequence, and M is matched to the length of P1. If the channel length is less than L_B samples, then the magnitude of the received preamble is periodic (except for the noise) and $AC[n]$ increases. Note that, thanks to the complex conjugation in Equation (5.13), this accumulation is relatively insensitive to phase rotation due to not yet compensated CFO. At the end of P1, the magnitude of $AC[n]$ stops increasing and this is used to detect the end of P1, hence the timing information is known. Several detectors have been proposed to locate this peak, but we will use a maximum function since Fort *et al.* (2003) suggest it is a good approximation of the maximum likelihood (ML) solution. Thus, the timing offset is estimated as follows:

$$\hat{n}_{AC,max} = \arg\max_n (AC[n]) \qquad (5.14)$$

GENERIC OFDM SYSTEM

Figure 5.11 shows the performance of the AC algorithms in a multi-path environment. The top histogram (noise only) provides no timing information (the peaks at each side are due to values beyond ±10) sampling instants. The other histograms show the distribution of the timing estimation at various SNR levels. The performance clearly improves with the SNR but it can be seen that, even at high SNR, the timing estimate needs to be refined.

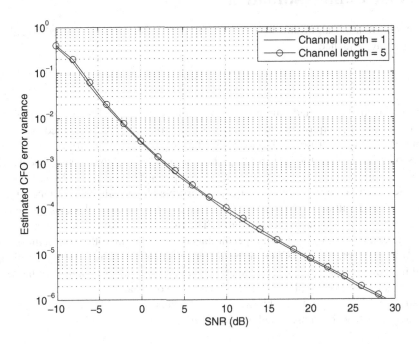

Figure 5.12 CFO estimation based on the auto-correlation of the received P1. The window length was equal to seven B periods out of 10, assuming three B periods were needed for burst detection and AGC estimation.

5.5 Coarse CFO estimation

Interestingly, AC provides a simple means to estimate the CFO as follows:

$$\Delta\omega = \frac{1}{2\pi L_B T} \angle(AC[\hat{n}_{AC,max}]) \tag{5.15}$$

where T is the baseband sample rate. The CFO can be estimated anywhere during the accumulation of the AC, but estimating it at $\hat{n}_{AC,max}$ obviously provides the best result. This estimator measures the phase rotation of the received signal over a duration $L_B T_S$. Note that the unambiguous estimation range of this estimator is equal to the inverse of the period in P1. Specifically, the unambiguous range is in the interval $](-1/2L_B T_S), (1/2L_B T_S)[=]-625\,\text{kHz}, 625\,\text{kHz}[$.[2] Figure 5.12 shows the error variance of the AC-based CFO estimation.

---

[2]Note that this is sufficient for 20 ppm offset at both the transmitter and receiver: $40 \times 10^{-6} \times 5 \times 10^{-9} = 232\,\text{kHz}$.

Notably, the variance is insensitive to the channel length (as long as it is shorter than the length of one B period). In Section 5.9, we also address CFO estimation in the presence of IQ imbalance.

5.6 Fine timing estimation

The cross-correlation algorithm is based on the correlation of the received signal with a known training sequence:

$$XC[n] = \sum_{m=0}^{L_C-1} p^*[m]r[n+m] \qquad (5.16)$$

where $p(m)$ is the transmitted training sequence of length $L_C = 64$ baseband samples. Note that this operation is equivalent to a matched filter for $p[m]$, as illustrated by the block diagram of the XC algorithm in Figure 5.13. The magnitude of the cross-correlator output consists of a large peak for each multi-path component, and several small peaks due to AWGN and imperfect auto-correlation properties of $p[m]$. The top plot of Figure 5.14 shows the cross-correlation output for the C sequence corrupted by a multi-path channel at an SNR of 3 dB, with and without CFO. Without CFO, the XC output resembles closely the channel impulse response, shown in the bottom curve. The frequency offset reduces significantly the size of the channel peaks, degrading performances. To minimize this problem, the XC is generally used after the coarse CFO has been compensated for; the residual CFO is then only a small fraction of the sub-carrier spacing. Clearly, in multi-path environments, the detection of the optimal timing is complicated. There is no single optimal detector for the peak because the optimal detector requires the knowledge of the channel profile. Several (heuristic) XC timing detectors have been presented by (Fort et al. 2003):

$$\text{XC-MAX} \quad \hat{n}_{XC-MAX} = \arg\max_n(|XC[n]|) \qquad (5.17)$$

$$\text{XC-SUM} \quad \hat{n}_{XC-SUM} = \arg\max_n\left(\sum_{p=1}^{L_{XC}} |XC[n+p]|\right) \qquad (5.18)$$

$$\text{XC-MIN} \quad \hat{n}_{XC-MIN} = \arg\min_n ||XC[n]| \geq th \times |XC(\hat{n}_{XC,max})| \qquad (5.19)$$

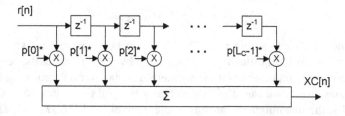

Figure 5.13 Cross-correlation block diagram.

GENERIC OFDM SYSTEM

The XC-MAX algorithm selects the index of the peak with the largest magnitude. The XC-SUM algorithm sums consecutive correlator outputs to locate a window of length L_{XC} where the channel has the most energy. L_{XC} is ideally equal to the (usually unknown) channel length L_C. The XC-MIN algorithm selects the earliest peak with a magnitude greater than some percentage of the largest peak. This method tends to select the first multi-path component rather than later reflections, thus reducing the variance of the timing estimate. The threshold must be chosen large enough to avoid selecting small noise peaks, but small enough to avoid selecting late multi-path peaks. The performance of the XC-SUM detector is shown in Figure 5.15. The SNR increases from top to bottom whereas the number of accumulated XC outputs is equal, from left to right, to L_C, $L_C + 3$ and $L_C + 6$. The performance is ideal in the left column because channel length is ideally known. *All* XC algorithms need some adjustment to modify the estimated timing so that the timing is moved a few samples to the left, because too early timing estimation consumes few samples of the CP but does not cause significant degradation; on the other hand, too late synchronization straddles over the next OFDM symbol, causing severe ISI.

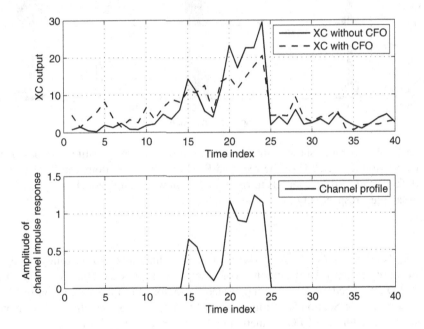

Figure 5.14 Cross-correlator output without CFO and with CFO equal to half the sub-carrier spacing (top). The degradation of the XC peak value is visible. The bottom curve shows the power profile of the channel used to generate the XC outputs of the top curve.

5.7 Fine CFO estimation

An auto-correlation on P2 can also be used to refine the CFO estimation. The repetition period within P1 is 3.2 µs; the unambiguous CFO estimation range lies in the interval

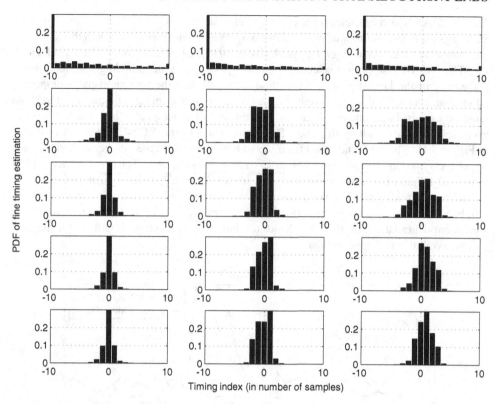

Figure 5.15 Histograms showing the performance of the timing estimation based on the cross-correlation (XC-SUM) of the received P2. From top to bottom, the SNR is equal to $-\infty$, 0, 5, 10 and 15 dB; from left to right, the window length is equal to L_C, $L_C + 3$ and $L_C + 6$ samples, respectively.

$](-1/2L_C T_S), (1/2L_C T_S)[=]-156\,\text{kHz}, 156\,\text{kHz}[$. This range is four times smaller than the acquisition range on P1 during the coarse acquisition. It is therefore important that the coarse CFO estimated on P1 be compensated for before the finer estimation on P2. The method described in Section 5.5 (Equations 5.13 to 5.15) can be directly applied. Note that Section 5.9 also addresses CFO estimation in the presence of IQ imbalance.

5.8 Complexity of auto- and cross-correlation

Another important issue with auto- and cross-correlation is complexity. The AC algorithm complexity is determined by its initialization and steady state complexities. At initialization, the first AC output requires M complex multiply-and-accumulate (CMAC) operations, where M is defined as in Equation (5.13). Then, for each new incoming sample, a new AC output is generated by one CMAC and one complex substraction (because $M - 1$ values are common to two neighboring AC outputs). For XC, there is no initialization complexity. At each incoming sample, L_C CMACs are consumed to generate one XC output. Therefore, the complexity of the XC is much higher than that of AC. For this reason – and according to the required performance – some XC algorithms are carried out on only a segment of the XC sequence.

5.9 Joint CFO and IQ imbalance acquisition

The frequency synchronization for OFDM systems is a well-studied problem in the scientific literature and the results are already integrated in recent wireless communication systems. Moose (1994) had first computed the ML estimator of the CFO in the frequency domain. Schmidl and Cox (1997) had shown that a periodic preamble could be used to estimate efficiently the CFO irrespective of the multi-path channel. The ML estimate of the CFO consists in assessing the received signal phase drift due to the CFO by correlating the repetitive parts of the received signal with each other (see Sections 5.5 and 5.7). The CFO estimation accuracy improves as the length of the preamble period increases since the observed phase drift becomes higher and is less subject to noise in that case. However, the unambiguous CFO range covered by the estimator decreases with the length of the preamble period since it is limited by a possible phase ambiguity. Therefore, the WLAN communication system based on the IEEE 802.11a/g standard proposes a two-step CFO estimation: a coarse estimation based on periodic short preamble sequences, followed by a fine estimation based on periodic long preamble sequences (van Nee *et al.* 1999).

Because the IQ imbalance degrades the estimation of the CFO (and reciprocally), the two effects need to be estimated together. Fouladifard and Shafiee (2003) and Yan *et al.* (2004) introduce two different estimators of the CFO in the presence of IQ imbalance. The estimator proposed by Fouladifard and Shafiee (2003) relies on a specifically designed pilot symbol in the frequency domain consisting of only one loaded carrier. The CFO and IQ imbalance are estimated iteratively to minimize the energy that they cause on the unloaded carriers. On the other hand, Yan *et al.* (2004) define a measure of the CFO with an expectation independent of the IQ imbalance and averages it over multiple short training sequences (compatible with the IEEE 802.11a/g periodic short preamble sequences). On the contrary, Tubbax *et al.* (2003) and Yu *et al.* (2005) propose an estimator of the IQ imbalance in the presence of CFO relying on a preamble composed of two identical long sequences (compatible with the IEEE 802.11a/g periodic long preamble sequences). It exploits the fact that IQ imbalance destroys the smoothness of the propagation channels in the frequency domain and deduces the IQ imbalance by observing the abrupt changes in the estimated frequency response. Interestingly, Xing *et al.* (2005) propose an integrated structure to compensate for the CFO and for frequency-dependent IQ imbalance. Again, the estimator relies on a specifically designed training sequence consisting of multiple identical short training sequences, where the even parts are rotated by a common phase. Finally, we have also proposed a new preamble structure based on which independent metrics of the CFO and IQ imbalance can be defined (Rore *et al.* 2006). The accuracy obtained with the corresponding estimator can fulfill the very strict specifications of the emerging 4G wireless systems.

In this section, we make use of the EM algorithm (see Moon (1996) for an introduction to the EM algorithm) to estimate iteratively the CFO and IQ imbalance (Horlin et al. 2007).[3] The solution relies on a preamble composed of two identical long sequences which enable the fine CFO acquisition independent of the channel knowledge. Any basic preamble sequence can be selected so that our system is compatible with the recent communication standards (for example, with the IEEE 802.11a/g periodic long preamble sequences). We target further a very high CFO and IQ imbalance estimation accuracy to meet the high specifications of emerging fourth generation (4G) wireless systems. In order to achieve a computationally tractable solution, the ML function is simplified by making a second-order approximation.

5.9.1 System model

We first derive the system model that will be used afterwards to compute the likelihood function.

Preamble description

The transmitted preamble $p[n]$ consists of a cyclic prefix of size L_{cp} at least equal to the channel impulse response length followed by a complex sequence $a[n]$ of length $Q \gg L_{cp}$ repeated one time ($a[n]$ is different from 0 for $n = 0, \ldots, Q-1$). The cyclic prefix is formed with the last samples of the sequence $a[n]$. The preamble sequence $p[n]$ is defined mathematically as:

$$p[n] := \begin{cases} a[n \bmod Q] & \text{if } n = -L_{cp}, \ldots, 2Q - 1 \\ 0 & \text{else} \end{cases} \quad (5.20)$$

Because the preamble is repetitive, the received vector should be repetitive also. However, this property is destructed by the front-end non-idealities (CFO and IQ imbalance).

Input–output relationship

In order to build the system model, we follow an approach similar to the one followed in Section 4.1.1. Compared to the model derived in Section 4.1.1, however, that integrates the CFO, SCO and IQ imbalance, we neglect the impact of the SCO since the acquisition has usually a small duration compared to the rest of the burst.

Due to the use of different local oscillators at the transmitter and at the receiver, the downconversion to the baseband domain is operated with a phase shift ϕ_0 and with a frequency shift $\Delta \omega$ (CFO). On the other hand, IQ imbalance caused by the use of different elements on the I and Q branches, is modeled by a difference in amplitude ϵ and phase $\Delta \phi$ between the two branches.

If $r(t) := r_r(t) + jr_i(t)$ is the baseband representation of the RF received signal $r_{RF}(t)$, the signal observed at the output of the analog front-end is given by (see the derivation of

---

[3] Portions reprinted, with permission by ©2007 IEEE from: François Horlin, André Bourdoux, Eduardo Lopez-Estraviz, Liesbet Van der Perre: 'Low-Complexity EM-based Joint Acquisition of the Carrier Frequency Offset and IQ imbalance', *IEEE Transactions on Wireless Communications*, accepted in September 2007.

GENERIC OFDM SYSTEM

Equations (4.10)):

$$y(t) = \alpha \, e^{-j(\Delta\omega t + \phi_0)} (r(t) \otimes \psi_R(t) \, e^{j\Delta\omega t})$$
$$+ \beta \, e^{j(\Delta\omega t + \phi_0)} (r(t) \otimes \psi_R(t) \, e^{j\Delta\omega t})^* \qquad (5.21)$$

where the parameters α and β are function of the IQ imbalance as defined in Equations (4.7) and (4.8), and $\psi_R(t)$ is the receiver analog front-end lowpass filter. The signal $y(t)$ is sampled at the rate $1/T$, leading to the received sequence $y[n] := y(t = nT)$.

Based on Equation (5.21), the received sequence can be expressed as a function of the transmitted preamble as:

$$y[n] = \alpha \, e^{-j(\Delta\omega nT + \phi_0)} (p[n] \otimes h[n]) + \beta \, e^{j(\Delta\omega nT + \phi_0)} (p[n] \otimes h[n])^* + z[n] \qquad (5.22)$$

in which $h[n]$ is the digital equivalent impulse response defined as:

$$h[n] := (\psi_T(t) \otimes c(t) \otimes \psi_R(t) \, e^{j\Delta\omega t})_{|t=NT} \qquad (5.23)$$

We assume that $h[n]$ has a finite length $L \leq L_{cp}$. The noise sequence $z[n]$ is obtained by observing $w_{RF}(t)$ at the output of the receiver analog front-end. Assuming that the filter $\psi_R(t)$ is a perfect lowpass filter of bandwidth $1/T$, the noise samples are independent and of variance $\sigma_z^2 = 2N_0/T(1+\epsilon^2) \simeq 2N_0/T$.

Vector model

After serial-to-parallel conversion, the received vector $\underline{y}$ can be decomposed into four parts: $\underline{y}_{cp}$ of size L_{cp} corresponding to the cyclic prefix, $\underline{y}_0$ and $\underline{y}_1$ of size Q corresponding to the repeated parts of the preamble, and $\bar{\underline{y}}$ of size $L-1$ caused by the convolution of the last samples of the preamble with the channel impulse response. By focusing on the parts $\underline{y}_0$ and $\underline{y}_1$ of the received vector given by:

$$\underline{y}_0 := [y[0] \cdots y[Q-1]]^T \qquad (5.24)$$
$$\underline{y}_1 := [y[Q] \cdots y[2Q-1]]^T \qquad (5.25)$$

and by defining the noise vectors as:

$$\underline{z}_0 := [z[0] \cdots z[Q-1]]^T \qquad (5.26)$$
$$\underline{z}_1 := [z[Q] \cdots z[2Q-1]]^T \qquad (5.27)$$

we obtain finally:

$$\underline{y}_0 = \underline{x} + \bar{\beta}\underline{x}^* + \underline{z}_0 \qquad (5.28)$$
$$\underline{y}_1 = e^{-j\phi}\underline{x} + \bar{\beta} \, e^{j\phi}\underline{x}^* + \underline{z}_1 \qquad (5.29)$$

in which $\underline{x} := [x[0] \cdots x[Q-1]]^T$ with $x[n] := \alpha \, e^{-j(\Delta\omega nT + \phi_0)}(p[n] \otimes h[n])$ is the reference received vector, $\phi := \Delta\omega QT$ is the phase difference between the two received signal parts due to the CFO, and $\bar{\beta} := \beta/\alpha^*$ is the modified IQ imbalance.

The ultimate aim is to exploit the redundancy in the received signal in order to build an estimate of the CFO and IQ imbalance independent of the propagation channel. It is therefore interesting to express the second part of the received signal $\underline{y}_1$ as a function of the first part $\underline{y}_0$. Based on Equation (5.28), it is clear that:

$$x = \frac{\underline{y}_0 - \bar{\beta}\underline{y}_0^*}{1 - |\bar{\beta}|^2} - \frac{\underline{z}_0 - \bar{\beta}\underline{z}_0^*}{1 - |\bar{\beta}|^2} \tag{5.30}$$

By introducing Equation (5.30) in Equation (5.29), we obtain:

$$\underline{y}_1 = \lambda(\phi, \bar{\beta})\underline{y}_0 + \mu(\phi, \bar{\beta})\underline{y}_0^* + \underline{n} \tag{5.31}$$

in which the parameters $\lambda(\phi, \bar{\beta})$ and $\mu(\phi, \bar{\beta})$ are defined as:

$$\lambda(\phi, \bar{\beta}) := \cos\phi - j\sin\phi \frac{1 + |\bar{\beta}|^2}{1 - |\bar{\beta}|^2} \tag{5.32}$$

$$\mu(\phi, \bar{\beta}) := 2j\sin\phi \frac{\bar{\beta}}{1 - |\bar{\beta}|^2} \tag{5.33}$$

and the noise vector $\underline{n}$ is given by:

$$\underline{n} := -\lambda(\phi, \bar{\beta})\underline{z}_0 - \mu(\phi, \bar{\beta})\underline{z}_0^* + \underline{z}_1 \tag{5.34}$$

The noise vector is formed with Gaussian distributed independent elements of zero mean. We will compute their variance in the next section.

Noise variance

The goal of this section is to compute the variance of the noise vector defined in Equation (5.34). Let $\underline{z}_{0r}$ and $\underline{z}_{0i}$ denote the real and imaginary parts of $\underline{z}_0$, $\underline{z}_{1r}$ and $\underline{z}_{1i}$ denote the real and imaginary parts of $\underline{z}_1$, and $\bar{\beta}_r$ and $\bar{\beta}_i$ denote the real and imaginary parts of $\bar{\beta}$. By taking the definitions (5.32) and (5.33) of the functions $\lambda(\phi, \bar{\beta})$ and $\mu(\phi, \bar{\beta})$ into account, the vector $\underline{n}$ is equal to:

$$\underline{n} = \underline{z}_{0r}\left[\left(-\cos\phi + 2\sin\phi\frac{\bar{\beta}_i}{1-|\bar{\beta}|^2}\right) + j\left(\sin\phi\frac{1+|\bar{\beta}|^2}{1-|\bar{\beta}|^2} - 2\sin\phi\frac{\bar{\beta}_r}{1-|\bar{\beta}|^2}\right)\right]$$

$$+ \underline{z}_{0i}\left[j\left(-\cos\phi - 2\sin\phi\frac{\bar{\beta}_i}{1-|\bar{\beta}|^2}\right) + \left(-\sin\phi\frac{1+|\bar{\beta}|^2}{1-|\bar{\beta}|^2} - 2\sin\phi\frac{\bar{\beta}_r}{1-|\bar{\beta}|^2}\right)\right]$$

$$+ \underline{z}_{1r} + j\underline{z}_{1i} \tag{5.35}$$

The variance of the real part $\underline{n}^R$ of $\underline{n}$ is equal to

$$\sigma_{nr}^2(\phi, \bar{\beta}) = \sigma_{z_{0r}}^2 \frac{1}{(1-|\bar{\beta}|^2)^2}(\cos\phi(1-|\bar{\beta}|^2) - 2\sin\phi\bar{\beta}_i)^2$$

$$+ \sigma_{z_{0i}}^2 \frac{1}{(1-|\bar{\beta}|^2)^2}(\sin\phi(1+|\bar{\beta}|^2) + 2\sin\phi\bar{\beta}_r)^2 + \sigma_{z_{1r}}^2 \tag{5.36}$$

GENERIC OFDM SYSTEM

and the variance of the imaginary part $\underline{n}^I$ of $\underline{n}$ is equal to

$$\sigma_{n_i}^2(\phi, \bar{\beta}) = \sigma_{z_{0r}}^2 \frac{1}{(1-|\bar{\beta}|^2)^2}(\sin\phi(1+|\bar{\beta}|^2) - 2\sin\phi\bar{\beta}_r)^2$$

$$+ \sigma_{z_{0i}}^2 \frac{1}{(1-|\bar{\beta}|^2)^2}(\cos\phi(1-|\bar{\beta}|^2) + 2\sin\phi\bar{\beta}_i)^2 + \sigma_{z_{1i}}^2 \quad (5.37)$$

in which $\sigma_{z_{0r}}^2 = \sigma_{z_{0i}}^2 = \sigma_{z_{1r}}^2 = \sigma_{z_{1i}}^2 = \sigma_z^2/2$ are the variances of the real and imaginary parts of the vectors $\underline{z}_0$ and $\underline{z}_1$, respectively. By adding the two terms (real and imaginary parts), the total variance of the elements of the noise $\underline{n}$ is equal to:

$$\sigma_n^2(\phi, \bar{\beta}) = \sigma_{n_r}^2(\phi, \bar{\beta}) + \sigma_{n_i}^2(\phi, \bar{\beta})$$

$$= \sigma_z^2 \left(2 - |\bar{\beta}|^2 + 8(\sin\phi)^2 \frac{|\bar{\beta}|^2}{1-|\bar{\beta}|^2}\right) \quad (5.38)$$

which shows that the variance of the noise samples depends on the CFO and IQ imbalance.

The following sections are devoted to the estimation of the phase ϕ and of the modified IQ imbalance $\bar{\beta}$.

5.9.2 Likelihood function and its second-order approximation

The ML estimate of the CFO and IQ imbalance is given by the set of parameters ϕ and $\bar{\beta}$ that maximizes the function:

$$\Lambda(\phi, \bar{\beta}) := \log p(\underline{y}_1|\underline{y}_0, \phi, \bar{\beta}) \quad (5.39)$$

$$= -\frac{Q}{2}\log(2\pi\sigma_n^2(\phi, \bar{\beta}))$$

$$- \frac{1}{2\sigma_n^2(\phi, \bar{\beta})}(\underline{y}_1 - \lambda(\phi, \bar{\beta})\underline{y}_0 - \mu(\phi, \bar{\beta})\underline{y}_0^*)^H \cdot (\underline{y}_1 - \lambda(\phi, \bar{\beta})\underline{y}_0 - \mu(\phi, \bar{\beta})\underline{y}_0^*) \quad (5.40)$$

where the last equality (5.40) has been obtained based on Equation (5.31) in which the noise vector has a Gaussian distribution.

Introducing the terms (5.32), (5.33) and (5.38) in the likelihood function (5.40) does not lead to a tractable function. For reasonable values of the SNR, the first term is negligible compared to the second term (σ_z^2 is small). On the other hand, because the IQ mismatch parameter $\bar{\beta}$ is small for actual analog front-ends ($\bar{\beta} \ll 1$), we can approximate the likelihood function by making a second-order approximation. The terms (5.32), (5.33) and (5.38) are approximately equal to:

$$\lambda(\phi, \bar{\beta}) \approx e^{-j\phi} - 2j\sin\phi|\bar{\beta}|^2 \quad (5.41)$$

$$\mu(\phi, \bar{\beta}) \approx 2j\sin\phi\bar{\beta} \quad (5.42)$$

$$\sigma_n^2(\phi, \bar{\beta}) \approx 2\sigma_v^2 \frac{1}{1 + (\frac{1}{2} - 4(\sin\phi)^2)|\bar{\beta}|^2} \quad (5.43)$$

Based on the last expressions and neglecting the terms of order higher than 2, the likelihood function (5.40) becomes:

$$\Lambda(\phi, \bar{\beta}) \approx \Lambda_0(\phi) + \Lambda_r(\phi, \bar{\beta}_r) + \Lambda_i(\phi, \bar{\beta}_i) \tag{5.44}$$

with:

$$\Lambda_0(\phi) := -\frac{1}{4\sigma_v^2} b_0(\phi) \tag{5.45}$$

$$\Lambda_r(\phi, \bar{\beta}_r) := -\frac{1}{4\sigma_v^2}(b_{1r}(\phi)\bar{\beta}_r + (b_2(\phi) + a_2(\phi)b_0(\phi))\bar{\beta}_r^2) \tag{5.46}$$

$$\Lambda_i(\phi, \bar{\beta}_i) := -\frac{1}{4\sigma_v^2}(b_{1i}(\phi)\bar{\beta}_i + (b_2(\phi) + a_2(\phi)b_0(\phi))\bar{\beta}_i^2) \tag{5.47}$$

in which $\bar{\beta}_r$ and $\bar{\beta}_i$ denote the real and imaginary parts of $\bar{\beta}$, and

$$a_2(\phi) := \tfrac{1}{2} - 4(\sin\phi)^2 \tag{5.48}$$

$$b_0(\phi) := (\underline{Z}_0^H \underline{Z}_0) + (\underline{Z}_1^H \underline{Z}_1) - 2\Re(\underline{Z}_1^H \underline{Z}_0)\cos\phi - 2\Im(\underline{Z}_1^H \underline{Z}_0)\sin\phi \tag{5.49}$$

$$b_{1r}(\phi) := -4\Im(\underline{Z}_0^T \underline{Z}_1)\sin\phi - 4\Re(\underline{Z}_0^T \underline{Z}_0)(\sin\phi)^2 + 4\Im(\underline{Z}_0^T \underline{Z}_0)\sin\phi\cos\phi \tag{5.50}$$

$$b_{1i}(\phi) := 4\Re(\underline{Z}_0^T \underline{Z}_1)\sin\phi - 4\Re(\underline{Z}_0^T \underline{Z}_0)\sin\phi\cos\phi - 4\Re(\underline{Z}_0^T \underline{Z}_0)(\sin\phi)^2 \tag{5.51}$$

$$b_2(\phi) := 8(\underline{Z}_0^H \underline{Z}_0)(\sin\phi)^2 - 4\Im(\underline{Z}_1^H \underline{Z}_0)\sin\phi \tag{5.52}$$

The term $a_2(\phi)$ is the term in $|\bar{\beta}|^2$ coming from the inverse of the noise variance in Equation (5.40). The term $b_0(\phi)$ is the term independent of $\bar{\beta}$, the terms $b_{1r}(\phi)$ and $b_{1i}(\phi)$ are the terms in $\bar{\beta}_r$ and $\bar{\beta}_i$ and the term $b_2(\phi)$ is the term in $|\bar{\beta}|^2$ coming from the vector product in Equation (5.40).

In the following sections, we show how the ML function (5.40) can be used to estimate the CFO when the system does not suffer from IQ imbalance and how the approximated ML function (5.44) can be used to estimate the CFO and the IQ imbalance iteratively based on the EM algorithm.

5.9.3 ML estimate of the CFO in the absence of IQ imbalance

When there is no IQ imbalance in the system ($\bar{\beta} = 0$), the ML function (5.40) reduces to:

$$\Lambda(\phi) := \log p(\underline{y}_1 | \underline{y}_0, \phi) \tag{5.53}$$

$$= -\frac{Q}{2}\log(4\pi\sigma_z^2) - \frac{1}{4\sigma_z^2}(\underline{y}_1 - e^{-j\phi}\underline{y}_0)^H \cdot (\underline{y}_1 - e^{-j\phi}\underline{y}_0) \tag{5.54}$$

Therefore, the ML estimate of the CFO is given by:

$$\hat{\phi} = \max_{\phi} \Lambda(\phi) \tag{5.55}$$

$$= \max_{\phi} \Re[\underline{y}_1^H \cdot \underline{y}_0 \, e^{-j\phi}] \tag{5.56}$$

$$= \angle(\underline{y}_1^H \cdot \underline{y}_0) \tag{5.57}$$

GENERIC OFDM SYSTEM

The ML estimate of the CFO in the absence of IQ imbalance is obtained by averaging the phase difference between the samples of the vectors $\underline{y}_0$ and $\underline{y}_1$ (auto-correlation method introduced by see Schmidl and Cox (1997) and presented in Sections 5.5 and 5.7). Even if this estimate is not accurate enough when the system suffers from IQ imbalance, it is an interesting starting point of the iterative algorithm proposed in the following section.

5.9.4 EM algorithm for the joint CFO and IQ imbalance estimation

A usual approach to maximize the likelihood function over two parameters is to estimate iteratively the parameters after each other. The iterative expectation maximization (EM) algorithm is known to converge to the joint ML estimate of the parameters. The EM algorithm is typically used to estimate a set of parameters based on a given observation when unknown random processes perturbate the observation. In our case, the phase ϕ due to the CFO is estimated based on the received vectors $\underline{y}_0$ and $\underline{y}_1$. The IQ imbalance $\bar{\beta}$ perturbates the estimation of ϕ.

Mathematically, one iteration of the EM algorithm is described as follows. If ϕ^i denotes the estimate of the phase ϕ at the iteration i, the estimate at the next iteration $i+1$ is given by (Georghiades and Han 1997; Moon 1996):

$$\phi^{i+1} = \arg\max_{\phi} \int_{\bar{\beta}_r} \int_{\bar{\beta}_i} \Lambda(\phi, \bar{\beta}) \, e^{\Lambda(\phi^i, \bar{\beta})} \, d\bar{\beta}_r \, d\bar{\beta}_i \qquad (5.58)$$

The phase selected at the new iteration is the one that maximizes the conditional expectation of the likelihood function over the IQ imbalance perturbation, given the most recent estimate of the phase.

Because the approximation (5.44) of the likelihood function (5.40) is decomposed into two independent functions (one depends on $\bar{\beta}_r$, the other depends on $\bar{\beta}_i$), it is also possible to simplify the double integral in Equation (5.58) in two independent integrals. The expression (5.44) is equal to:

$$\phi^{i+1} = \arg\max_{\phi} (-b_0(\phi) - b_{1r}(\phi)\bar{\beta}_r^i - b_{1i}(\phi)\bar{\beta}_i^i - (b_2(\phi) + a_2(\phi)b_0(\phi))(\bar{\beta}_r^2)^i$$
$$- (b_2(\phi) + a_2(\phi)b_0(\phi))(\bar{\beta}_i^2)^i) \qquad (5.59)$$

in which the following definitions have been made:

$$\bar{\beta}_r^i := \int_{\bar{\beta}_r} \bar{\beta}_r F_r(\bar{\beta}_r) \, d\bar{\beta}_r \qquad (5.60)$$

$$\bar{\beta}_i^i := \int_{\bar{\beta}_i} \bar{\beta}_i F_i(\bar{\beta}_i) \, d\bar{\beta}_i \qquad (5.61)$$

$$(\bar{\beta}_r^2)^i := \int_{\bar{\beta}_r} (\bar{\beta}_r^2) F_r(\bar{\beta}_r) \, d\bar{\beta}_r \qquad (5.62)$$

$$(\bar{\beta}_i^2)^i := \int_{\bar{\beta}_i} (\bar{\beta}_i^2) F_i(\bar{\beta}_i) \, d\bar{\beta}_i \qquad (5.63)$$

with:

$$F_r(\bar{\beta}_r) := \frac{e^{\Lambda_r(\phi^i, \bar{\beta}_r)}}{\int_{\bar{\beta}_r} e^{\Lambda_r(\phi^i, \bar{\beta}_r)} \, d\bar{\beta}_r} \quad (5.64)$$

$$F_i(\bar{\beta}_i) := \frac{e^{\Lambda_i(\phi^i, \bar{\beta}_i)}}{\int_{\bar{\beta}_i} e^{\Lambda_i(\phi^i, \bar{\beta}_i)} \, d\bar{\beta}_i} \quad (5.65)$$

Interestingly, the function $F_r(\bar{\beta}_r)$ can be seen as the probability distribution of a Gaussian random variable of mean $-b_{1r}(\phi^i)/2(b_2(\phi^i) + a_2(\phi^i)b_0(\phi^i))$ and of variance $2\sigma_v^2/(b_2(\phi^i) + a_2(\phi^i)b_0(\phi^i))$. Equivalently, the function $F_i(\bar{\beta}_i)$ can be seen as the probability distribution of a Gaussian random variable of mean $-b_{1i}(\phi^i)/2(b_2(\phi^i) + a_2(\phi^i)b_0(\phi^i))$ and of variance $2\sigma_v^2/(b_2(\phi^i) + a_2(\phi^i)b_0(\phi^i))$. Therefore, we obtain:

$$\bar{\beta}_r^i = -\frac{b_{1r}(\phi^i)}{2(b_2(\phi^i) + a_2(\phi^i)b_0(\phi^i))} \quad (5.66)$$

$$\bar{\beta}_i^i = -\frac{b_{1i}(\phi^i)}{2(b_2(\phi^i) + a_2(\phi^i)b_0(\phi^i))} \quad (5.67)$$

$$(\bar{\beta}_r^2)^i = \frac{2\sigma_v^2}{(b_2(\phi^i) + a_2(\phi^i)b_0(\phi^i))} + (\bar{\beta}_r^i)^2 \quad (5.68)$$

$$(\bar{\beta}_i^2)^i = \frac{2\sigma_v^2}{(b_2(\phi^i) + a_2(\phi^i)b_0(\phi^i))} + (\bar{\beta}_i^i)^2 \quad (5.69)$$

The two last equalities have been obtained because the auto-correlation is equal to the variance plus the square of the mean. It is interesting to note that the expressions (5.66) and (5.67) correspond to the ML estimate of the parameters $\bar{\beta}_r$ and $\bar{\beta}_i$ for a given phase ϕ equal to ϕ^i (the mean of a Gaussian variable is also the value that maximizes the distribution). The estimate of the square of the parameters $\bar{\beta}_r$ and $\bar{\beta}_i$ given in Equations (5.68) and (5.69) is, however, equal to the square of the ML estimates plus a correction term.

The EM iterative algorithm can be summarized as follows.

1. Start from an initial estimate of the phase ϕ^0.

2. At the iteration i, evaluate the parameters $a_2(\phi^i)$, $b_0(\phi^i)$, $b_{1r}(\phi^i)$, $b_{1i}(\phi^i)$ and $b_2(\phi^i)$ according to the definitions (5.48) to (5.52) based on the value ϕ^i.

3. Estimate the real part $\bar{\beta}_r^i$ and the imaginary part $\bar{\beta}_i^i$ of the IQ imbalance, and their square $(\bar{\beta}_r^2)^i$ and $(\bar{\beta}_i^2)^i$ according to the results (5.66) to (5.69).

4. Select the phase ϕ^{i+1} such that the function (5.59) is maximized.

5. Iterate on the steps 2 to 4, until the algorithm converges.

The initial phase estimate is typically obtained by neglecting the IQ imbalance and by using the algorithm proposed by Schmidl and Cox (1997). A block diagram of the acquisition system is provided is Figure 5.16.

GENERIC OFDM SYSTEM

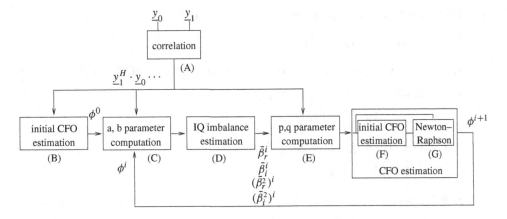

Figure 5.16 System block diagram.

5.9.5 Implementation

The critical part of the computational effort resides in the search of the value ϕ^{i+1} that maximizes the function (5.59) at each iteration (step 4 in the previous section). Obviously, the parameters (5.48) to (5.52) and the function (5.59) can be computed for all possible values of ϕ, and the optimal value ϕ^{i+1} can be selected. In this section, we simplify rather the search of the maximum based on trigonometric identities and based on the method of Newton–Raphson to find the roots of a function.

Assuming that the sine and cosine functions are implemented with a lookup table, it is interesting to apply the following trigonometric identities in order to avoid the multiplications:

$$(\sin \phi)^2 = \tfrac{1}{2} - \tfrac{1}{2} \cos (2\phi) \tag{5.70}$$

$$\sin \phi \cos \phi = \tfrac{1}{2} \sin (2\phi) \tag{5.71}$$

$$(\sin \phi)^3 = \tfrac{3}{4} \sin \phi - \tfrac{1}{4} \sin (3\phi) \tag{5.72}$$

$$(\sin \phi)^2 \cos \phi = \tfrac{1}{4} \cos \phi - \tfrac{1}{4} \cos (3\phi) \tag{5.73}$$

The equalities (5.48) to (5.52) can be easily modified according to Equations (5.70) to (5.73) and included in the expression (5.59) of the phase estimate. It gives:

$$\phi^{i+1} = \arg\max_{\phi} G(\phi) \tag{5.74}$$

with:

$$G(\phi) := p_1(\bar{\beta}^i) \cos \phi + q_1(\bar{\beta}^i) \sin \phi + p_2(\bar{\beta}^i) \cos (2\phi) + q_2(\bar{\beta}^i) \sin (2\phi) \\ + p_3(\bar{\beta}^i) \cos (3\phi) + q_3(\bar{\beta}^i) \sin (3\phi) \tag{5.75}$$

in which $\bar{\beta}^i$ denotes the set of parameters $\{\bar{\beta}_r^i, \bar{\beta}_i^i, (\bar{\beta}_r^2)^i, (\bar{\beta}_i^2)^i\}$, and:

$$p_1(\bar{\beta}^i) := \Re(\underline{Z}_1^H \underline{Z}_0) (2 - (\bar{\beta}_r^2)^i - (\bar{\beta}_i^2)^i) \tag{5.76}$$

$$q_1(\bar{\beta}^i) := \Im(\underline{Z}_1^H \underline{Z}_0) (2 - (\bar{\beta}_r^2)^i - (\bar{\beta}_i^2)^i) + 4(\Im(\underline{Z}_0^T \underline{Z}_1) \bar{\beta}_r^i - \Re(\underline{Z}_0^T \underline{Z}_1) \bar{\beta}_i^i) \tag{5.77}$$

$$p_2(\bar{\beta}^i) := -2 \, (\Re(\underline{Z}_0^T \underline{Z}_0) \bar{\beta}_r^i + \Im(\underline{Z}_0^T \underline{Z}_0) \bar{\beta}_i^i)$$
$$+ 2((\bar{\beta}_r^2)^i + (\bar{\beta}_i^2)^i) ((\underline{Z}_0^H \underline{Z}_0) - (\underline{Z}_1^H \underline{Z}_1)) \tag{5.78}$$

$$q_2(\bar{\beta}^i) := -2 \, (\Im(\underline{Z}_0^T \underline{Z}_0) \bar{\beta}_r^i - \Re(\underline{Z}_0^T \underline{Z}_0) \bar{\beta}_i^i) \tag{5.79}$$

$$p_3(\bar{\beta}^i) := 2 \Re(\underline{Z}_1^H \underline{Z}_0) ((\bar{\beta}_r^2)^i + (\bar{\beta}_i^2)^i) \tag{5.80}$$

$$q_3(\bar{\beta}^i) := 2 \Im(\underline{Z}_1^H \underline{Z}_0) ((\bar{\beta}_r^2)^i + (\bar{\beta}_i^2)^i) \tag{5.81}$$

Finding the maximum in Equation (5.75) is equivalent to finding the root of the first-order derivative $G'(\phi)$ of the function $G(\phi)$. It gives:

$$G'(\phi^{i+1}) = 0 \tag{5.82}$$

with:

$$G'(\phi) := -p_1(\bar{\beta}^i) \sin \phi + q_1(\bar{\beta}^i) \cos \phi - 2p_2(\bar{\beta}^i) \sin(2\phi)$$
$$+ 2q_2(\bar{\beta}^i) \cos(2\phi) - 3p_3(\bar{\beta}^i) \sin(3\phi) + 3q_3(\bar{\beta}^i) \cos(3\phi) \tag{5.83}$$

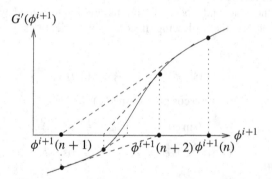

Figure 5.17 The Newton–Raphson method.

We propose to apply the iterative method of Newton–Raphson to evaluate the roots of a function (see Figure 5.17). The new value is given by the intersection with the x-axis of the tangent of the function at the current value. At the iteration n, the new estimate of the root is given by:

$$\phi^{i+1}(n+1) = \phi^{i+1}(n) - \frac{G'(\phi^{i+1}(n))}{G''(\phi^{i+1}(n))} \tag{5.84}$$

GENERIC OFDM SYSTEM

in which $G''(\phi^{i+1}(n))$ denotes the second order derivative of $G(\phi)$. In order to have a reliable initial estimate, it is useful to remark that the terms in $p_1(\bar{\beta}^i)$ and $q_1(\bar{\beta}^i)$ in Equation (5.83) are dominant. A good starting point is therefore:

$$\phi^{i+1}(0) = \begin{cases} \arctan\left(\dfrac{q_1(\bar{\beta}^i)}{p_1(\bar{\beta}^i)}\right) & \text{if } p_1 \geq 0 \\ \arctan\left(\dfrac{q_1(\bar{\beta}^i)}{p_1(\bar{\beta}^i)}\right) + \pi & \text{if } p_1 < 0, q_1 \geq 0 \\ \arctan\left(\dfrac{q_1(\bar{\beta}^i)}{p_1(\bar{\beta}^i)}\right) - \pi & \text{if } p_1 < 0, q_1 < 0 \end{cases} \quad (5.85)$$

because $\phi^{i+1}(0)$ is a root of the function $G'(\phi)$ corresponding to a maximum of the function $G(\phi)$ (and not to a minimum) if the terms $p_2(\bar{\beta}^i)$, $q_2(\bar{\beta}^i)$, $p_3(\bar{\beta}^i)$ and $q_3(\bar{\beta}^i)$ are negligible.

5.9.6 Performance and complexity analysis

The goal of this section is to compare thoroughly the proposed EM algorithm to the state-of-the-art CFO/IQ estimator proposed by Tubbax et al. (2003) in terms of performance and complexity. Both systems rely on a possibly standard compliant repeated long training sequence, that enables a fine estimation of the parameters.

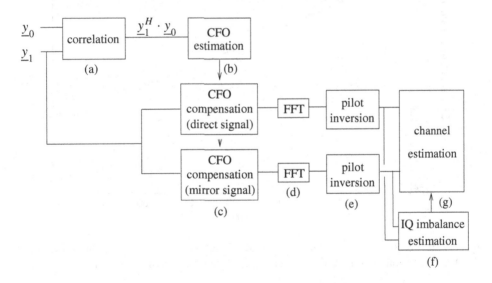

Figure 5.18 State-of-the art CFO/IQ imbalance estimator.

The principle of the state-of-the-art acquisition scheme is summarized in the block diagram provided in Figure 5.18. The CFO is first estimated by observing the phase drift between the two received preamble sequences (correlation and CFO estimation blocks).

A correction of the CFO is applied independently on the direct received signal and on the mirror received signal due to IQ imbalance. After FFT, the channel can be estimated in the frequency domain by inverting the pilot symbols on each carrier. The IQ imbalance parameter is chosen that minimizes the abrupt changes in the estimated channel response.

In order to assess the performance and complexity of both schemes, we consider a typical WLAN communication system operating in an indoor environment at 5 GHz with a 20 MHz bandwidth. According to the IEEE 802.11a standard, each transmitted burst consists of a preamble followed by regular OFDM symbols. The first 4 µs of the preamble are used for burst detection and AGC. The next 4 µs of the preamble can be used for packet detection, and coarse timing/frequency synchronization. The last 8 µs of the preamble are used for synchronization refinement and channel estimation. We focus on the last part of the preamble dedicated to the fine synchronization and composed of a cyclic prefix of length 32, followed by a sequence of length 64 (the C sequence) repeated one time. The performance is averaged over 1000 channel realizations.

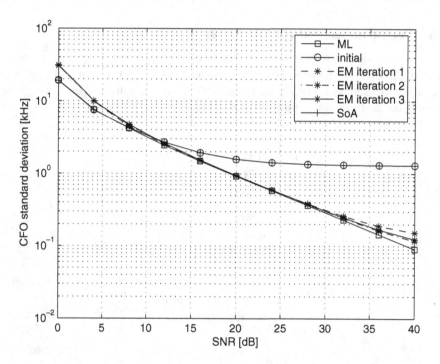

Figure 5.19 Frequency estimate error standard deviation as a function of the SNR, $\phi = 2$, $\bar{\beta} = 0.1\, e^{j3\pi/8}$. Note that the SoA estimate is equivalent to the initial estimate.

The standard deviation of the estimated frequency error and of the estimated IQ imbalance error is illustrated in Figures 5.19 and 5.20, respectively, for the different estimation schemes as a function of the SNR. We assume a fixed value of the CFO equal to 50 kHz ($\phi = 2$) and a fixed value of the IQ imbalance equal to 10% ($\bar{\beta} = 0.1\, e^{j3\pi/8}$). The initial CFO estimator (Schmidl and Cox) has a performance similar to the ML algorithm at low

GENERIC OFDM SYSTEM

SNR and floors to a poor estimate at high SNR because it does not take the IQ imbalance into account. The state-of-the-art acquisition scheme proposed by Tubbax *et al.* (2003) builds on the initial CFO estimate to evaluate further the IQ imbalance. Starting from the initial CFO estimate, the EM algorithm converges progressively to the CFO and IQ imbalance ML estimates and outperforms significantly the state-of-the-art acquisition scheme. Fewer than three iterations are needed to reach the point of convergence. At low SNR (less than 10 dB), the EM algorithm suffers, however, from the fact that we have neglected the first term in Equation (5.40) to obtain the approximated likelihood function (5.44). At high SNR (more than 35 dB), the EM algorithm suffers also from the second-order approximation in the IQ imbalance performed to obtain the approximated likelihood function (5.44). Comparing Figures 5.19 and 5.20, it is clear that mostly the EM estimate of the IQ imbalance suffers from the two approximations while the EM estimate of the CFO is quite robust.

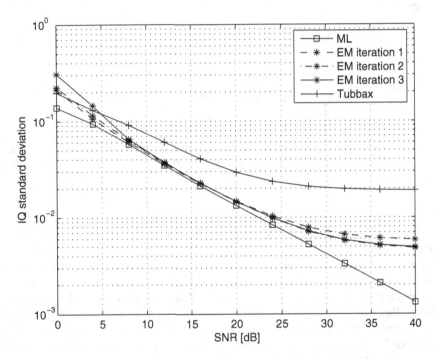

Figure 5.20 IQ imbalance estimation error standard deviation as a function of the SNR, $\phi = 2$, $\bar{\beta} = 0.1\, e^{j3\pi/8}$.

Figure 5.21 illustrates the standard deviation of the frequency estimation error (solid curves) and of the IQ imbalance estimation error (dashed curves) as a function of the IQ imbalance for an SNR equal to 20 dB. The performance of the initial frequency estimator degrades rapidly with the value of the IQ imbalance, while the EM frequency and IQ imbalance estimator performance is nearly independent from the value of the IQ imbalance. The performance of the EM algorithm remains close to the one of the ML algorithm. Even if

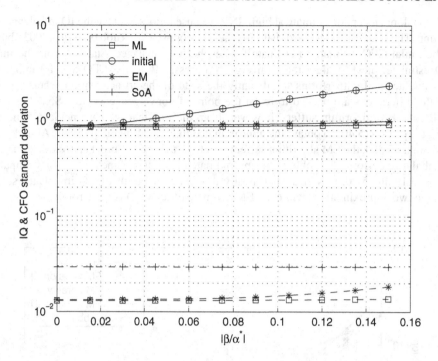

Figure 5.21 CFO and IQ imbalance error standard deviation as a function of the IQ imbalance. Solid curves: CFO; dashed curves: IQ imbalance, $\phi = 2$, $\angle \bar{\beta} = 3\pi/8$, SNR = 20 dB.

the IQ imbalance state-of-the-art estimate is stable, it is always significantly less accurate than the EM estimate.

Figure 5.22 illustrates the standard deviation of the frequency estimation error (solid curves) and of the IQ imbalance estimation error (dashed curves) as a function of the CFO for an SNR equal to 20 dB. The performance of the EM algorithm is close to the one of the ML algorithm. The quality of the CFO estimate varies slightly as a function of the CFO (except on the borders of the range where there is a phase ambiguity between $+\pi$ and $-\pi$). On the other hand, the quality of the IQ imbalance estimate is strongly dependent on the CFO. When $\phi = 0$ (CFO = 0) or when $\phi = \pm\pi$ (CF0 = ± 150 kHz), the two vectors $\underline{Z}_0$ and $\underline{Z}_1$ in the model (5.28) to (5.29) become identical (except for the contribution of the noise). It is therefore impossible to discriminate between the two unknown parameters $\bar{\beta}$ and $\underline{X}$ based on the observation of $\underline{Z}_0$ and $\underline{Z}_1$. In order to solve this problem, an artificial CFO can be added on the same preamble sent a second time as proposed by Rore et al. (2006). Another solution would be to use the state-of-the-art IQ imbalance estimator which is globally worse but more stable when the CFO varies.

Tables 5.2 and 5.3 give the complexity of the EM algorithm and of the state-of-the-art CFO/IQ imbalance estimator proposed by Tubbax et al. (2003) for each constituting block (see Figures 5.16 and 5.18). The complexity is estimated in the number of real operations (addition, multiplication or division) and in the number of accesses to the lookup tables

GENERIC OFDM SYSTEM

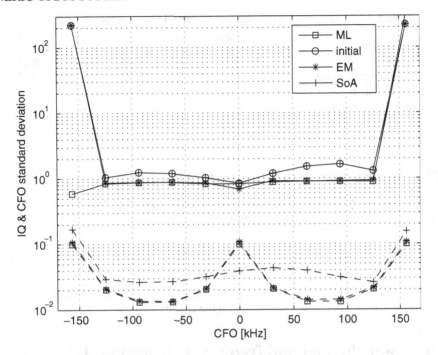

Figure 5.22 CFO and IQ imbalance error standard deviation as a function of the CFO. Solid curves: CFO; dashed curves: IQ imbalance, $\phi = 2$, $\angle \bar{\beta} = 3\pi/8$, SNR = 20 dB.

Table 5.2 Complexity of the EM algorithm.

Block	(A)	(B)	(C)	(D)	(E)	(F)	(G)	Total
Typical occurrence	1	1	2	2	2	2	4	
Real additions	$20Q$	—	40	5	12	—	49	1590
Real multiplications	$20Q$	—	10	1	11	—	12	1372
Real divisions	—	1	—	3	—	1	1	13
Accesses to arctan	—	1	—	—	—	1	—	3
Accesses to cos	—	—	4	—	—	—	6	32

(arctangent or cosine). To evaluate the complexity of the EM algorithm, we have observed that two EM iterations and two Newton–Raphson iterations per EM iteration are sufficient in practice. The complexity of the EM algorithm is significantly lower than that of the CFO/IQ imbalance estimator proposed by Tubbax et al. (2003). The main reason is that the EM algorithm iterates over a few real variables while the state-of-the-art solution processes blocks of complex samples over the whole chain.

Table 5.3 Complexity of the state-of-the-art CFO/IQ imbalance estimator.

Block	(a)	(b)	(c)	(d)	(e)	(f)	Total
Real additions	$4Q$	—	$4Q$	$8Q \log Q$	—	$16Q + 3$	4611
Real multiplications	$4Q$	—	$4Q$	$8Q \log Q$	$2Q$	$16Q + 6$	4742
Real divisions	—	1	—	—	—	2	3
Accesses to arctan	—	1	—	—	—	—	1
Accesses to cos	—	—	$2Q$	—	—	—	128

5.9.7 Compensation of the CFO and IQ imbalance

Once both effects have been estimated, they can be successively compensated for by processing the received signal $y[n]$ (Equation (5.22)) as follows:

$$[y[n] - \bar{\beta} y^*[n]] e^{jn(\phi/Q)} = \frac{|\alpha|^2 - |\beta|^2}{\alpha^*} e^{-j\phi} (p[n] \otimes h[n]) \tag{5.86}$$

The constant factor in front of the desired term can be seen as a part of the channel response (it is estimated and compensated together with the channel response).

5.10 Joint channel and frequency-dependent IQ imbalance estimation

In the previous section, we have shown that the frequency-independent IQ imbalance generated in the local oscillators can be reliably estimated together with the CFO during the synchronization phase, and compensated afterwards in the time domain. Frequency-dependent IQ imbalance generated in the receiver amplifiers and filters should rather be estimated and compensated in the frequency domain in order to keep the low complexity properties of the OFDM system.

Different schemes have been proposed in the literature to deal with the frequency-dependent IQ imbalance. The contributions of Valkama *et al.* (2001) and Pun *et al.* (2001) propose a first compensation structure of the frequency-dependent IQ imbalance based on the uncorrect assumption that the signals on the I and Q branches are uncorrelated. Xing *et al.* (2005) develop a joint estimator of the CFO and frequency-dependent IQ imbalance in the time domain that suffers unfortunately from an unacceptable computational complexity. Ylamurto (2003) introduces a new method to estimate the frequency-dependent IQ imbalance together with the frequency selective channel that relies on an asymmetrical pilot symbol to cancel the interference caused by the IQ imbalance in the frequency domain. Schuchert *et al.* (2001) demonstrate that the frequency-dependent IQ imbalance can also be compensated after the channel equalization in the frequency domain by applying a two-tap filter on each sub-carrier independently. Finally Jian and Tsui (2004) propose an adaptive frequency-dependent IQ imbalance estimation/compensation structure in the frequency domain working per carrier that is suboptimal because it does not exploit the channel frequency correlation.

The estimation and compensation structure considered in this book is near to the one introduced recently by Lopez-Estraviz *et al.* (2006) and Lopez-Estraviz *et al.* (2007).

GENERIC OFDM SYSTEM

We propose to estimate the frequency-dependent IQ imbalance during the channel estimation phase and to compensate for it during the channel equalization phase.

5.10.1 System model

When the system suffers from frequency-dependent IQ imbalance, the transmitted signal is received through two different paths: the direct path $h_\alpha[n]$ and the mirror path $h_\beta[n]$ (see the equality Equation (4.41) for the demonstration). The channel equalization is based on the knowledge of the two corresponding impulse responses that need therefore to be estimated.

A model of the channel estimation system is illustrated in Figure 5.23. The joint estimation of the impulse responses $h_\alpha[n]$ and $h_\beta[n]$ relies on the transmission of the OFDM pilot symbol block $\tilde{p}$ of size Q in the frequency domain. After IFFT, cyclic prefix addition and parallel-to-serial conversion, the signal is sent through the direct channel $h_\alpha[n]$ and through the mirror channel $h_\beta[n]$. After serial-to-parallel conversion, removal of the cyclic prefix and FFT, the received vector in the frequency domain $\tilde{y}$ is equal to (see the Equation (4.45) in which the CFO and SCO have been neglected):

$$\tilde{\underline{y}} = \underline{\underline{\Lambda}}_{\tilde{p}} \cdot \tilde{\underline{h}}_\alpha + \underline{\tilde{I}}_Q \cdot \underline{\underline{\Lambda}}_{\tilde{p}}^* \cdot \tilde{\underline{h}}_\beta^* + \tilde{\underline{z}} \qquad (5.87)$$

where $\underline{\underline{\Lambda}}_{\tilde{p}}$ is a diagonal matrix of size Q containing the pilot symbol on its diagonal, $\tilde{\underline{h}}_\alpha$ and $\tilde{\underline{h}}_\beta$ are vectors of size Q composed of the channel coefficients in the frequency domain, and $\tilde{\underline{z}}$ is the noise in the frequency domain.

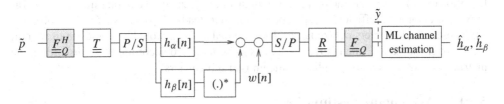

Figure 5.23 Channel estimation in the presence of the frequency-dependent IQ imbalance.

The conventional channel estimator, in the absence of IQ imbalance, consists in evaluating each channel coefficient independently in the frequency domain by observing the corresponding pilot symbol at the receiver. The complexity of the estimator is low as it acts on a per-sub-carrier basis. However, this method is sub-optimal because it does not exploit the correlation between the frequency domain channel coefficients caused by the limited channel time duration. Furthermore it is impossible to estimate jointly the $2Q$ channel coefficients contained in the vectors $\tilde{\underline{h}}_\alpha$ and $\tilde{\underline{h}}_\beta$ based on the observation $\tilde{\underline{y}}$ of size Q. Therefore, the channels should rather be estimated in the time domain where their finite time duration can be exploited to decrease the number of parameters to estimate. We can assume without loss of generality that the channels are of order equal to the cyclic prefix length L_{cp} and define the channel vectors in the time domain as:

$$\underline{h}_\alpha := [h_\alpha[0] \cdots h_\alpha[L_{cp}-1]]^T \tag{5.88}$$

$$\underline{h}_\beta := [h_\beta[0] \cdots h_\beta[L_{cp}-1]]^T \tag{5.89}$$

The time domain vectors are function of the frequency domain vectors as expressed in:

$$\underline{\tilde{h}}_\alpha = \underline{\underline{F}} \cdot \underline{h}_\alpha \tag{5.90}$$

$$\underline{\tilde{h}}_\beta = \underline{\underline{F}} \cdot \underline{h}_\beta \tag{5.91}$$

where $\underline{\underline{F}}$ is a matrix of size $Q \times L_q$ composed of the L_q first columns of the FFT matrix of size Q. Taking Equations (5.90) and (5.91) into account, the initial expression (5.87) can be written as:

$$\underline{\tilde{y}} = \underline{\underline{\Delta}}_{\tilde{p}} \cdot \underline{\underline{F}} \cdot \underline{h}_\alpha + \underline{\underline{I}}_Q \cdot \underline{\underline{\Delta}}_{\tilde{p}}^* \cdot \underline{\underline{F}}^* \cdot \underline{h}_\beta^* + \underline{\tilde{z}} \tag{5.92}$$

or, by taking a matrix notation, it can be synthetically written as:

$$\underline{\tilde{y}} = \underline{\underline{P}} \cdot \underline{h} + \underline{\tilde{z}} \tag{5.93}$$

where the channel vector is defined as:

$$\underline{h} := \begin{bmatrix} \underline{h}_\alpha^T & \underline{h}_\beta^H \end{bmatrix}^T \tag{5.94}$$

and the pilot symbol matrix is defined as:

$$\underline{\underline{P}} := \begin{bmatrix} \underline{\underline{\Delta}}_{\tilde{p}} \cdot \underline{\underline{F}} & \underline{\underline{I}}_Q \cdot \underline{\underline{\Delta}}_{\tilde{p}}^* \cdot \underline{\underline{F}}^* \end{bmatrix} \tag{5.95}$$

It is important to note the interest of estimating the time domain channels $\underline{h}_\alpha$ and $\underline{h}_\beta$ based on Equation (5.92) rather than the frequency domain ones $\underline{\tilde{h}}_\alpha$ and $\underline{\tilde{h}}_\beta$ based on Equation (5.87) as only $2L_{cp}$ complex coefficients need to be estimated instead of $2Q$ ($L_{cp} \ll Q$). Note also that any subset of sub-carriers of size larger than or equal to $2L_{cp}$ would be sufficient to support the channel estimation (the matrix model has to be modified accordingly by discarding the matrix rows/columns corresponding to the unused sub-carriers).

5.10.2 ML channel estimation

The channel estimator is optimally designed according to the ML criterion. Assuming that the noise vector elements are independent of variance σ_z^2, the linear ML estimate of the channel vector $\underline{h}$ is given by (see Appendix B):

$$\underline{\hat{h}} = (\underline{\underline{P}}^H \cdot \underline{\underline{P}})^{-1} \cdot \underline{\underline{P}}^H \cdot \underline{\tilde{y}} \tag{5.96}$$

$$= \underline{h} + (\underline{\underline{P}}^H \cdot \underline{\underline{P}})^{-1} \cdot \underline{\underline{P}}^H \cdot \underline{\tilde{z}} \tag{5.97}$$

where it can be seen that the number of observed sub-carriers must be at least equal to $2L_{cp}$ so that the inner product $\underline{\underline{P}}^H \cdot \underline{\underline{P}}$ is full rank and can be inverted. The channel estimation error auto-correlation matrix at the output of the ML estimator is:

$$\underline{\underline{R}}_{h_\epsilon h_\epsilon} = \mathcal{E}[(\underline{h} - \underline{\hat{h}}) \cdot (\underline{h} - \underline{\hat{h}})^H] \tag{5.98}$$

$$= \sigma_z^2 (\underline{\underline{P}}^H \cdot \underline{\underline{P}})^{-1} \tag{5.99}$$

GENERIC OFDM SYSTEM

Taking the Equation (5.95) of the pilot symbol matrix $\underline{\underline{P}}$ into account and assuming that the pilot symbols are on the unit circle, the expression (5.99) reduces to:

$$\underline{\underline{R}}_{h_\epsilon h_\epsilon} = \sigma_z^2 \left[\begin{array}{cc} Q\underline{\underline{I}}_L & \underline{\bar{F}}^H \cdot \underline{\underline{\Delta}}_{\tilde{p}}^* \cdot \underline{\tilde{\underline{I}}}_Q \cdot \underline{\underline{\Delta}}_{\tilde{p}}^* \cdot \underline{\bar{F}}^* \\ \underline{\bar{F}}^T \cdot \underline{\underline{\Delta}}_{\tilde{p}} \cdot \underline{\tilde{\underline{I}}}_Q \cdot \underline{\underline{\Delta}}_{\tilde{p}} \cdot \underline{\bar{F}} & Q\underline{\underline{I}}_L \end{array} \right]^{-1} \quad (5.100)$$

The error variance of each channel coefficient estimate stands on the corresponding element on the diagonal of $\underline{\underline{R}}_{h_\epsilon h_\epsilon}$.

It has been recently shown by Lopez-Estraviz *et al.* (2006) and Lopez-Estraviz *et al.* (2007) that the pilot symbols can be selected such that the out-of-diagonal elements of the error auto-correlation matrix are null:

$$\underline{\bar{F}}^T \cdot \underline{\underline{\Delta}}_{\tilde{p}} \cdot \underline{\tilde{\underline{I}}}_Q \cdot \underline{\underline{\Delta}}_{\tilde{p}} \cdot \underline{\bar{F}} = \underline{\underline{0}}_L \quad (5.101)$$

In that case, the estimation error variance is minimized and the joint ML channel estimation of the impulse responses $\underline{h}_\alpha$ and $\underline{h}_\beta$ reduces to the separate ML estimation of the two impulse responses. The pilot symbols must satisfy the following condition:

$$\angle(p^q) + \angle(p^{-q}) = \pi q \quad (5.102)$$

for $q = -Q/2, \ldots, Q/2 - 1$. The degrees of freedom remaining in the selection of the sequence can be used to decrease the PAPR of the corresponding transmitted signal.

5.10.3 Performance and complexity analysis

The goal of this section is to assess the performance and the complexity of the proposed channel estimator when two different types of pilot symbol are used:

- the IEEE 802a standard compliant pilot symbol, dedicated to both fine synchronization and channel estimation (C sequence of length 64);
- any pilot symbol satisfying the optimality condition (5.102).

The achievable performance depends only on the transmitted pilot symbol, and not on the propagation channel or on the IQ imbalance. Following the WLAN system parameters, we assume OFDM blocks of 64 carriers and channel impulse responses of maximum order 16.

The standard deviation of the channel coefficient estimation error is illustrated in Figure 5.24 as a function of the carrier index for an SNR equal to 0 dB ($\sigma_z^2 = 1$ in the expression (5.100)). The channel vectors $\underline{\tilde{h}}_\alpha$ and $\underline{\tilde{h}}_\beta$ are estimated with the same error variance, so that only one result is shown for the two frequency responses. When the IEEE 802.11a pilot symbol is transmitted, the error standard deviation depends heavily on the carrier index. When the optimal pilot symbol is transmitted, the error standard deviation is independent on the carrier index. Using an optimal pilot symbol block brings a significant performance improvement.

The standard deviation of the channel coefficient estimation error averaged over the set of carriers is illustrated in Figure 5.25 as a function of the SNR. As shown in the expression (5.100), the channel estimation error variance depends linearly on the variance of the noise,

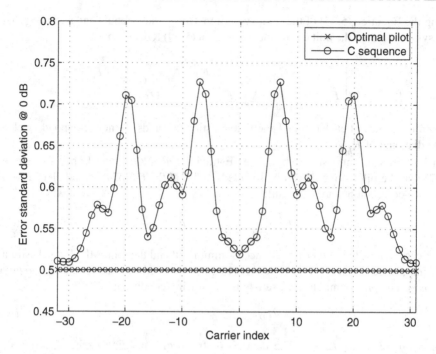

Figure 5.24 Channel estimation error standard deviation ($\underline{\tilde{h}}_\alpha$ and $\underline{\tilde{h}}_\beta$).

explaining the linearly decreasing curves. The gain achieved by the use of an optimal pilot symbol is also observed.

The complexity of the channel estimator is given in Table 5.4 for the general case and in Table 5.5 for the optimal pilot symbol case. We assume that the ML estimator can be precomputed as it depends only on the known pilot symbol so that the associated complexity can be neglected. When the pilot symbol is optimal, the complexity of the channel estimator is significantly lower since the joint estimation of the two channel impulse responses reduces to the separate estimation of each of them that can be performed at a low complexity based on FFT/IFFT operations.

5.10.4 Frequency-dependent IQ imbalance compensation

When an OFDM symbol block $\underline{\tilde{x}}$ is transmitted, the received vector in the frequency domain is equal to (see the model (4.45) in which the CFO and SCO have been neglected):

$$\underline{\tilde{y}} = \underline{\underline{\Lambda}}_{h_\alpha} \cdot \underline{\tilde{x}} + \underline{\underline{\tilde{I}}}_Q \cdot \underline{\underline{\Lambda}}^*_{h_\beta} \cdot \underline{\tilde{x}}^* + \underline{\tilde{z}} \qquad (5.103)$$

The ideal OFDM system can be described as a set of parallel independent sub-channels (first term of Equation (5.103)). In the presence of IQ imbalance, however, each sub-channel is corrupted by interference generated by another sub-channel located on the mirror position of the OFDM block (second term of Equation (5.103)).

GENERIC OFDM SYSTEM

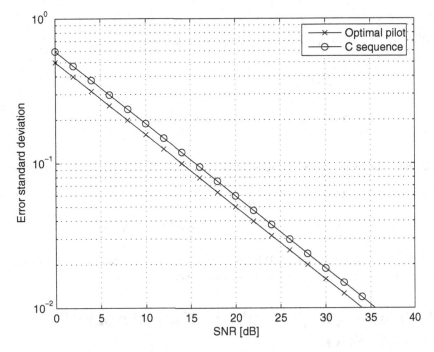

Figure 5.25 Channel estimation error standard deviation ($\tilde{\underline{h}}_\alpha$ and $\tilde{\underline{h}}_\beta$).

Table 5.4 Complexity of the channel estimator in the general case.

	Matrix product	FFT	Total
Real additions	$8LQ^2$	$8Q \log_2 Q$	$\sim 530k$
Real multiplications	$8LQ^2$	$8Q \log_2 Q$	$\sim 530k$

Table 5.5 Complexity of the channel estimator when the pilot symbol is optimal.

	Pilot inversion	IFFT	FFT	Total
Real additions	Q	$8Q \log_2 Q$	$8Q \log_2 Q$	$\sim 6k$
Real multiplications	—	$8Q \log_2 Q$	$8Q \log_2 Q$	$\sim 6k$

When the channel frequency responses $\tilde{\underline{h}}_\alpha$ and $\tilde{\underline{h}}_\beta$ are known, the interference generated by the IQ imbalance can be eliminated in the frequency domain by combining each received

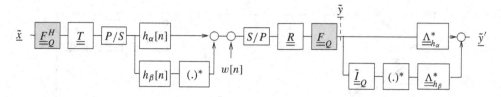

Figure 5.26 Frequency-dependent IQ imbalance compensation.

sub-channel with the complex conjugate of the interfering sub-channel (see Figure 5.26):

$$\underline{\tilde{y}}' = \underline{\Lambda}^*_{h_\alpha} \cdot \underline{\tilde{y}} + \underline{\Lambda}^*_{h_\beta} \cdot \underline{\tilde{I}}_Q \cdot \underline{\tilde{y}}^* \quad (5.104)$$

$$= (\underline{\Lambda}^*_{h_\alpha} \cdot \underline{\Lambda}_{h_\alpha} + \underline{\Lambda}^*_{h_\beta} \cdot \underline{\Lambda}_{h_\beta}) \cdot \underline{\tilde{x}} + \underline{\Lambda}^*_{h_\alpha} \cdot \underline{\tilde{z}} + \underline{\Lambda}^*_{h_\beta} \cdot \underline{\tilde{I}}_Q \cdot \underline{\tilde{z}}^* \quad (5.105)$$

A conventional zero-forcing or MMSE linear channel equalizer can be implemented afterwards to obtain the final estimate of the transmitted vector of symbols.

5.11 Tracking loops for phase noise and residual CFO/SCO

After the timing offset and CFO have been acquired and the channel coefficients (and I/Q imbalance) have been estimated (see all previous sections of this chapter), the processing of the OFDM symbols containing the payload data can start. It is usual, especially at low SNR, for some residual CFO error to remain after acquisition. If nothing is done to estimate and compensate this residual CFO, significant BER degradation may result. It is also likely that some SCO is present since the acquisition only compensated for a time offset. Finally, some phase noise is usually present which also requires some compensation; phase noise is a multiplicative process, so it also impacts the reception at high SNR. A practical way to cope with these three problems is to use a phase-tracking loop to compensate for the phase noise and residual CFO and a timing tracking loop to compensate for the SCO. We will refer to these loops as the CFO tracking loop and SCO tracking loop in the rest of this chapter.

5.11.1 Estimation of residual CFO/SCO

We consider an OFDM system where N_P pilots symbols are multiplexed in the frequency domain with the data symbols. At the OFDM symbol index n, we collect the transmitted pilot symbols in vector $\tilde{p}[n]$ and the received pilot symbols in vector $\tilde{y}[n]$. Neglecting ICI and the small amplitude perturbations, we have the following relationship

$$\tilde{y}^{q_P}[n] = \tilde{\phi}^{q_P}[n]\tilde{h}^{q_P}[n]\tilde{p}^{q_P}[n] + \tilde{w}^{q_P}[n] \quad (5.106)$$

where n is the OFDM symbol index, $q_P \in 1, \ldots, N_P$ is the pilot index and $\tilde{\phi}^{q_P}[n]$ is the phase rotation defined in Equation (4.21) due to CFO and SCO. To derive the estimators of the CFO and SCO, we follow an approach inspired from Oberli (2007). An approximate

GENERIC OFDM SYSTEM

log-likelihood function $\Lambda(\tilde{y} \mid \Delta\omega, \delta)$ can be derived from Equation (5.106) leading to:

$$\Lambda(\tilde{y} \mid \Delta\omega, \delta) = \frac{1}{2N_0 Q T} \sum_{p=1}^{N_P} \{(\tilde{y}^{q_P})^* \tilde{h}^{q_P} \tilde{p}^{q_P} e^{j\tilde{\phi}^{q_P}(\Delta\omega,\delta)}$$

$$+ \tilde{y}^{q_P} (\tilde{h}^{q_P})^* (\tilde{p}^{q_P})^* e^{-j\tilde{\phi}^{q_P}(\Delta\omega,\delta)}\} + C \qquad (5.107)$$

where C does not depend on $\Delta\omega$ or δ. Note that we do not show the dependence on n in Equation (5.107) for readability and that we used $\tilde{\phi}^{q_P}(\Delta\omega, \delta)$ to highlight the dependence of $\tilde{\phi}$ on $\Delta\omega$ and δ. Solving for $\Delta\omega$ or δ in Equation (5.107) implies taking the gradients with respect to $\Delta\omega$ or δ and setting those gradients to zero. These calculations are lengthy and we just summarize the approach. A system of two equations with two unknowns results, which does not have a closed-form solution. A work-around consists in solving first for $\Delta\omega$, assuming that δ is 0. This is justified by the fact that the CFO is usually the dominant effect. The resulting ML CFO estimator is then found to be:

$$\widehat{\Delta\omega}[n] = \frac{1}{2\pi Q} \angle \left(\sum_{p=1}^{N_P} (\tilde{y}^{q_P}[n])^* \tilde{h}^{q_P}[n] \tilde{p}^{q_P}[n] \right) \qquad (5.108)$$

which is exact (neglecting ICI) if $\delta = 0$. Note that the effect of CFO in the frequency domain is to rotate the phases of all sub-carriers by the same amount (ICI is also generated). With the knowledge of the estimated CFO $\widehat{\Delta\omega}$, one equation remains to be solved to find the estimated SCO $\hat{\delta}$. This again does not have a closed-form solution and an ad hoc solution consists in estimating the SCO as follows:

$$\hat{\delta}[n] = \frac{1}{\pi N_P (N_P - 1)} \sum_{p_1=1}^{N_P} \sum_{p_2=1}^{N_P} \frac{\angle \left(\tilde{e}^{q_{p_1}}[n] (\tilde{e}^{q_{p_2}}[n])^* \right)}{q_{p_1} - q_{p_2}} \qquad (5.109)$$

where

$$\tilde{e}^q[n] = (\tilde{y}^q[n])^* \tilde{h}^q[n] \tilde{p}^q[n] \qquad (5.110)$$

The estimator in Equation (5.109) actually averages all phase differences between all possible pairs of pilots. This can be shown to be equivalent to finding the slope of the phase shifts across the sub-carriers by linear regression.

Now that we are equipped with CFO and SCO estimators operating on one OFDM symbol, we can develop the tracking loops that will estimate and track the CFO and SCO on a sequence of received OFDM symbols. We will base the development of the tracking loop on the extended analysis provided in Gardner (2005).

5.11.2 CFO tracking loop

The block diagram of the CFO tracking loop is illustrated in Figure 5.27. The top part of the figure shows the components of the tracking loop in the receiver block diagram. The CFO compensation block consists of a complex mixer that shifts the received signal by a frequency offset equal to the estimated CFO. Then, the cyclic prefix is removed and the signal is converted to the frequency domain. The frequency offset is estimated by means of equation (5.108). The estimated CFO is then filtered by a loop filter to generate a smoothed estimate of the CFO, which is then used to generate the samples of a complex exponential used in the complex mixer. This process actually compensates an OFDM symbol by a filtered estimate obtained from the previous OFDM symbols. Hence, the *current* OFDM symbol must still be rotated in a feedforward fashion by the current CFO estimate. The equivalent linearized control loop model of the tracking loop is shown in the bottom of Figure 5.27. Two closed loop functions are interesting:

$$H_1(z) = \frac{\hat{\phi}(z)}{\phi(z)} = \frac{F(z)}{1+F(z)} \qquad (5.111)$$

$$H_2(z) = \frac{\phi_{err}(z)}{\phi(z)} = \frac{1}{1+F(z)} \qquad (5.112)$$

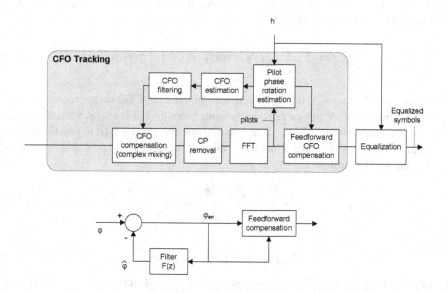

Figure 5.27 CFO tracking loop in the OFDM receiver (top) and equivalent linearized control loop model (bottom).

This is the classical phase-locked loop problem: the residual phase ϕ_{err} must be forced close to zero whereas the estimated phase $\hat{\phi}$ must follow the input phase as closely as possible. A suitable filter to achieve this goal is a filter of the form:

$$\hat{\phi}[n] = \hat{\phi}[n-1] + K_\phi(\phi_{err}[n] - \alpha\phi_{err}[n-1]) \qquad (5.113)$$

which corresponds to the z-transform:

$$F(z) = \frac{\hat{\phi}(z)}{\phi_{err}(z)} = K_\phi \frac{1 - \alpha z^{-1}}{1 - z^{-1}} \qquad (5.114)$$

GENERIC OFDM SYSTEM

In this filter, K_ϕ is the filter gain and α is a filter coefficient. They control the transient behavior of the filter. Note that $F(z)$ includes an integrator (the term $1 - z^{-1}$ in the denominator), which ensures that the residual error of the filter is 0 in the steady state. The closed loop transfer functions $H_1(z)$ and $H_2(z)$ become:

$$H_1(z) = \frac{\hat{\phi}(z)}{\phi(z)} = \frac{K_\phi * (1 - \alpha z_{-1})}{K_\phi + 1 - (1+\alpha)z^{-1}} \qquad (5.115)$$

$$H_2(z) = \frac{\phi_{err}(z)}{\phi(z)} = \frac{1 - z_{-1}}{K_\phi + 1 - (1+\alpha)z^{-1}} \qquad (5.116)$$

Equation (5.115) can be used to analyze the transient behavior and the stability of the loop. In addition, the choice of K_ϕ and α are application-dependent. They depend on system parameters such as the residual (i.e. after acquisition) CFO to be tracked and the characterization of the phase noise. For the current system under consideration and a practical level of phase noise, suitable values for K_ϕ and α are 0.8 and 0.9, respectively. Note also that this model is ideal since it does not take the ICI effect into account, which will cause degradation with respect to the ideal, linearized performance.

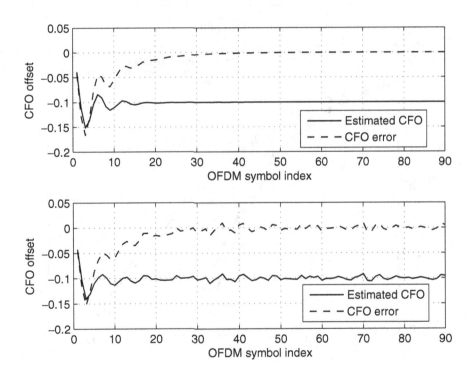

Figure 5.28 CFO tracking loop response in presence of CFO only (top) and CFO with phase noise (bottom). The CFO is equal to 10% of the sub-carrier spacing. The phase noise parameters are: -25 dBc integrated phase noise, corner frequency 100 kHz, phase noise floor -130 dBc/Hz.

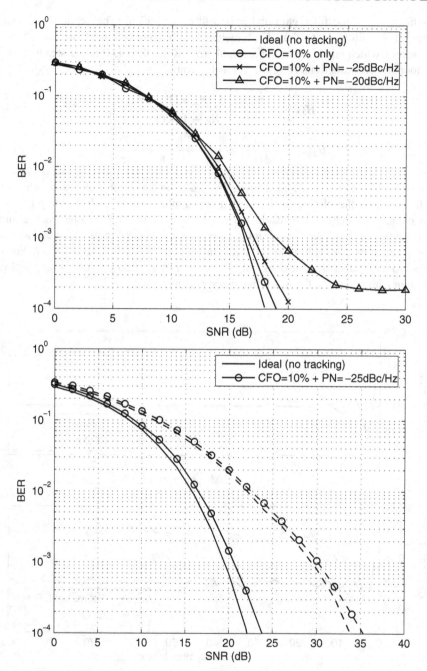

Figure 5.29 BER performance of the OFDM system in the presence of CFO and phase noise, in an AWGN channel (top) and in frequency selective channels (bottom). Low-delay spread channels: solid line; high-delay spread channels: dashed line.

GENERIC OFDM SYSTEM

The behavior of the CFO tracking loop is shown in Figure 5.28, without additive noise. The top figure shows the loop response to CFO only, with a CFO equal to 10% of the sub-carrier spacing. It can be seen that the loop steady state error is zero. The bottom figure shows the loop response to combined CFO offset and phase noise (phase noise parameters: -25 dBc integrated phase noise, corner frequency 100 kHz, phase noise floor -130 dBc/Hz). Obviously, the phase noise prevents to obtain an exact estimate of the CFO but the loop filtering effect allows to maintain the phase error close to zero. Figure 5.29 (top) shows the BER performance of the considered OFDM system with the same amount of residual CFO and phase noise as used for Figure 5.28 in an AWGN channel. We observe only a minor degradation even in the presence of combined CFO and phase noise levels up to -25 dBc/Hz. At phase noise levels of -20 dBc/Hz, a flooring begins to appear. Finally, Figure 5.29 (bottom) shows the performance in frequency selective channels, both for low- and high-delay spread channels. Low-delay spread channels are more affected because, in high-delay spread channels, the severe multi-path is responsible for the relatively higher degradation, hence masking the degradation due to the CFO and phase noise.

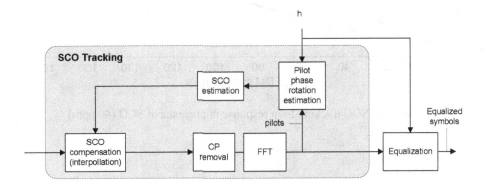

Figure 5.30 SCO tracking loop in the OFDM receiver.

5.11.3 SCO tracking loop

The block diagram of the SCO tracking loop is illustrated in Figure 5.30, showing the components of the tracking loop in the receiver block diagram. The SCO compensation block consists of a time-domain interpolator that resamples digitally (see Equation (3.54)) the received signal so that the symbol rate in the receiver is (ideally) equal to the one at the transmitter. Then, the cyclic prefix is removed and the signal is converted to the frequency domain. The sampling clock offset is estimated by means of Equation (5.109). The estimated SCO is then filtered by a loop filter to generate a smoothed estimate of the SCO, which is then used to generate the sampling times for the interpolator. This process actually compensates an OFDM symbol by a filtered SCO estimate obtained from the previous OFDM symbols. It is not necessary to compensate the SCO on the current symbol because the degradation is

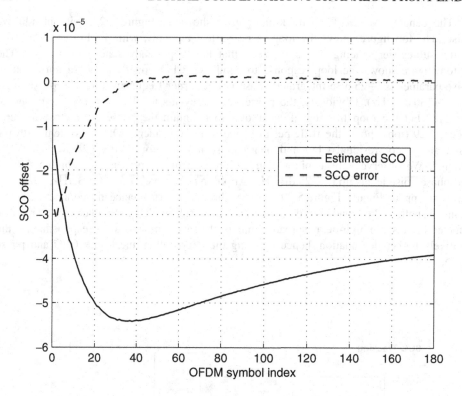

Figure 5.31 SCO tracking loop response in presence of SCO (40 ppm).

marginal (SCO is only a problem if it is uncompensated on a long sequence of symbols). The equivalent linearized control loop model of the SCO tracking loop is similar to that introduced for CFO tracking (see previous sub-section). The same filter structure as in Equations (5.111) and (5.114) is used to smooth the measured sample clock offset. For the current system under consideration and a practical level of phase noise, suitable values for K_ϕ and α in the SCO tracking loop are 0.5 and 0.75, respectively. Note again that this model is ideal since it does not take the ICI effect into account, which will cause degradation with respect to the ideal, linearized performance.

The behavior of the SCO tracking loop is shown in Figure 5.31, without additive noise, for an SCO of 40 ppm (4×10^{-5}). It can be observed that the loop steady state error is zero. Figure 5.32 (top) shows the BER performance of the considered OFDM system with a receiver SCO of 40 ppm, in an AWGN channel. We observe some degradation. Finally, Figure 5.32 (bottom) shows the performance in frequency selective channels, both for low- and high-delay spread channels. Just as for CFO, low-delay spread channels are more affected because, in high-delay spread channels, the severe multi-path is responsible for the relatively higher degradation, hence masking the degradation due to the SCO.

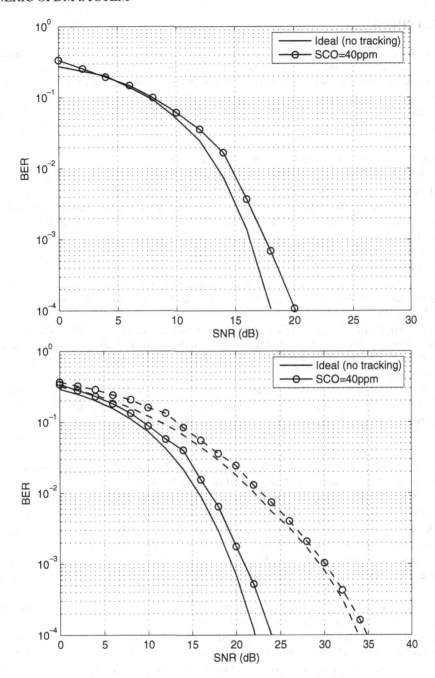

Figure 5.32 BER performance of the OFDM system in the presence of SCO (40 ppm), in an AWGN channel (top) and in frequency selective channels (bottom). Low-delay spread channels: solid line; high-delay spread channels: dashed line.

References

Fort, A., Weijers, J., Derudder, V., Eberle, W. and Bourdoux, A. (2003) A performance and complexity comparison of auto-correlation and cross-correlation for OFDM burst synchronization. *IEEE Proceedings of International Conference on Acoustics, Speech and Signal Processing*, pp. II341–II344.

Fouladifard, S. and Shafiee, H. (2003) Frequency offset estimation in OFDM systems in presence of IQ imbalance. *IEEE Proceedings of International Conference on Communications*, vol. 3, pp. 2071–2075.

Gardner, F. (2005) *Phaselock Techniques, 3rd edn.* Wiley-Interscience.

Georghiades, C. and Han, J. (1997) Sequence estimation in the presence of random parameters via the EM algorithm. *IEEE Transactions on Communications* **45**(3), 300–3008.

Hazy, L. and El-Tanany, M. (1997) Synchronization of OFDM systems over frequency selective fading channels. *IEEE Proceedings of Vehicular Technology Conference*, vol. 3, pp. 2094–2098.

Horlin, F., Bourdoux, A., Lopez-Estraviz, E. and Van der Perre, L. (2007) Low-complexity EM-based joint CFO and IQ imbalance acquisition. *IEEE Proceedings of International Conference on Communications*, pp. 2871–2876.

Jian, L. and Tsui, E. (2004) Joint adaptive transmitter/receiver IQ imbalance correction for OFDM systems. *IEEE Proceedings of Personal, Indoor and Mobile Radio Communications*, vol. 2, pp. 1511–1516.

Lopez-Estraviz, E., De Rore, S., Horlin, F., Bourdoux, A. and Van der Perre, L. (2007) Pilot design for joint channel and frequency-dependent transmit/receive IQ imbalance estimation and compensation in OFDM-based transceivers. *IEEE Proceedings of International Conference on Communications*, pp. 4861–4866.

Lopez-Estraviz, E., De Rore, S., Horlin, F. and Van der Perre, L. (2006) Optimal training sequences for joint channel and frequency-dependent IQ imbalance estimation in OFDM-based receivers. *IEEE Proceedings of International Conference on Communications*, vol. 10, pp. 4595–4600.

Moon, T. (1996) The expectation maximization algorithm. *IEEE Signal Processing Magazine* **13**(6), 47–59.

Moose, P. (1994) A technique for orthogonal frequency division multiplexing frequency offset correction. *IEEE Transactions on Communications* **42**(2), 2908–2914.

Oberli, C. (2007) ML-based tracking algorithms for MIMO-OFDM. *IEEE Transactions on Wireless Communications*.

Pun, K., Franca, J., Aceredo-Leme, C., Chan, C. and Choy, C. (2001) Correction of frequency-dependent I/Q mismatches in quadrature receivers. *IEEE Electronic Letters* **37**(23), 1415–1417.

De Rore, S., Lopez-Estraviz, E., Horlin, F. and Van der Perre, L. (2006) Joint estimation of carrier frequency offset and IQ imbalance for 4G mobile wireless systems. *IEEE Proceedings of International Conference on Communications*, vol. 5, pp. 2066–2071.

Schmidl, T. and Cox, D. (1997) Robust frequency and timing synchronization for OFDM. *IEEE Transactions on Communications* **45**(12), 1613–1621.

Schuchert, A., Hashlozner, R. and Antoine, P. (2001) A novel IQ imbalance compensation scheme for the reception of OFDM signals. *IEEE Transactions on Consumer Electronics* **47**(3), 313–318.

Tubbax, J., Come, B., Van der Perre, L., Deneire, L., Donnay, S. and Engels, M. (2003) Compensation of IQ imbalance in OFDM systems. *IEEE Proceedings of International Conference on Communications*, vol. 5, pp. 3403–3407.

Tufvesson, F., Edfors, O. and Faulkner, M. (1999) Time and frequency synchronization for OFDM using PN-sequence preambles. *IEEE Proceedings of Vehicular Technology Conference*, vol. 4, pp. 2203–2207.

Valkama, M., Renfors, M. and Koivunen, V. (2001) Compensation of frequency-selective I/Q imbalances in wideband receivers: models and algorithms. *IEEE Proceedings of Signal Processing Advances in Wireless Communications*, pp. 42–45.

van Nee, R., Awater, G., Morikura, M., Takanashi, H., Webster, M. and Halford, K. (1999) New high-rate wireless LAN standards. *IEEE Communications Magazine* **37**(12), 82–88.

Xing, G., Shen, M. and Liu, H. (2005) Frequency offset and I/Q imbalance compensation for direct-conversion receivers. *IEEE Transactions on Wireless Communications* **4**(2), 673–680.

Yan, F., Zhu, W. and Ahmad, M.O. (2004) Carrier frequency offset estimation for OFDM systems with I/Q imbalance. *IEEE Proceedings of 47th International Midwest Symposium on Circuits and Systems*, vol. 2, pp. 633–636.

Ylamurto, T. (2003) Frequency domain IQ imbalance correction scheme for orthogonal frequency division multiplexing (OFDM) systems. *IEEE Proceedings of Wireless Communications and Networking Conference*, vol. 1, pp. 20–25.

Yu, J., Sun, M., Hsu, T. and Lee, C. (2005) A novel technique for I/Q imbalance and CFO compensation in OFDM systems. *IEEE Proceedings of the International Symposium on Circuits and Systems*, vol. 6, pp. 6030–6033.

6

Emerging Wireless Communication Systems

At the beginning of the book, we introduced the air interfaces emerging in the new communication systems and evaluated their performance. We presented the possible analog front-end architectures and highlighted different sources of possible non-ideality. The degradation caused by the non-ideal front-end on the performance of the communication systems was also assessed. A generic approach to compensate numerically for the front-end non-ideality was developed. It is composed of two distinct phases:

- the acquisition, taking care of the rough estimation/compensation of the effects and relying on a preamble sent at the beginning of the burst;

- the tracking, taking care of the fine estimation/compensation of the effects and relying on pilot symbols interleaved in the data symbols.

In this final chapter, we show that the methodology developed to cope with the non-ideal front-end can be extended and applied to the most recent communication systems. We will successively address the IEEE 802.11n WLAN MIMO communication system (Section 6.1) and the 3GPP LTE multi-user cellular communication system (Section 6.2). Our goal is not to solve the problem completely for each emerging communication system, but rather to address the most challenging issues resulting from the particular configuration of each system. For each system, we will present the status of the standard and the targeted objectives, we will describe the burst organization (location of preamble and pilot symbols...), we will identify the major challenges to implement the system, and we will finally provide a solution for the most specific challenges.

Digital Compensation for Analog Front-Ends   François Horlin and André Bourdoux
© 2008 John Wiley & Sons, Ltd

6.1 IEEE 802.11n

6.1.1 Context

History

The WLAN PHY and MAC layers are standardized in the frame of the IEEE 802 LAN/MAN Standards Committee.[1] The specific working group dealing with WLANs is the IEEE802.11 Wireless LAN Working Group.[2] This working group has been active since the end of the 1990s and has produced WLAN specifications in use today such as the IEEE802.11, IEEE802.11b, IEEE802.11a and IEEE802.11g. More advanced systems (IEEE802.11n and IEEE802.11-VHT) are being developed. A brief history of the PHY evolution of the IEEE802.11 standard family is as follows (see also Table 6.1):

Table 6.1 Evolution of WLAN standards.

IEEE Standard	Modulation	Bandwidth	Frequency band	Peak bit rate	Timing
IEEE802.11	FHSS, 2/4-level GFSK	1 MHz	2.4 GHz	2 Mbps	1999
	DSSS, DBPSK, DQPSK	22 MHz	2.4 GHz	2 Mbps	1999
IEEE802.11b	DSSS, 8-chip CCK	22 MHz	2.4 GHz	11 Mbps	1999
IEEE802.11a	OFDM 2/4 PSK, 16/64 QAM	20 MHz	5 GHz	54 Mbps	1999
IEEE802.11g	OFDM 2/4 PSK, 16/64 QAM	20 MHz	2.4 GHz	54 Mbps	2003
IEEE802.11n	MIMO-OFDM 2/4 PSK, 16/64 QAM	20/40 MHz	2.4/5 GHz	600 Mbps	2009?
IEEE802.11-VHT	TBD	TBD	TBD	> 1 Gbps	2011?

- IEEE802.11:
  The original version of the IEEE 802.11 (IEEE802.11 1999) specified two different PHY for use in the 2.4 GHz ISM band: FH-SS and DS-SS. Two rates of 1 and 2 Mbps were defined for both the FHSS and the DSSS. In FHSS, the modulation was 2-level GFSK for 1 Mbps and 4-level GFSK for 2 MBps. In DSSS, the modulation was DBPSK for 1 Mbps and 4-level GFSK for 2 MBps. The chip rate for the DSSS system was 11 MHz, resulting in a RF bandwidth of about 22 MHz. The IEEE802.11 version was first released in 1997.

- IEEE802.11b:
  This amendment (IEEE802-11b 1999) to IEEE802.11 built on the IEEE802 data rate capabilities to provide 5.5 Mbps and 11 Mbps payload data rates in addition to the 1 Mbps and 2 Mbps rates. To provide the higher rates, 8-chip CCK is employed as the modulation scheme. The chipping rate remained at 11 MHz, which is the same as the

---
[1] URL of the IEEE 802 LAN/MAN Standards Committee: http://www.ieee802.org/.
[2] URL of the IEEE802.11 Wireless LAN Working Group: http://www.ieee802.org/11/.

DSSS system, thus providing the same occupied channel bandwidth. The IEEE802.11b version was first released in 1999.

- IEEE802.11a:
  This amendment (IEEE802.11a 1999) to IEEE802.11 aimed at exploiting the allocation of new bandwidth at 5 GHz for ISM applications. OFDM was identified as the technology of choice for the PHY, thanks to its elegant and efficient way of mitigating multi-path distortion. An OFDM system with 64 sub-carriers and a bandwidth of 20 MHz was defined (very similar to the 'generic OFDM system' described in Chapter 5). Various MCS were defined with bit rates ranging from 6 to 54 Mbps. The IEEE802.11a version was first released in 1999 but its commercialization began significantly later, mainly because the semiconductor technology was not sufficiently mature for these rates and carrier frequency.

- IEEE802.11g:
  This amendment (IEEE802.11g 2003) to IEEE802.11 aimed at reusing the OFDM PHY of the IEEE802.11a in the 2.4 GHz band. The IEEE802.11 version was released in 2003. It contributed largely to the deployment of commercial WLAN products using OFDM because of the lower operating frequency compared to the IEEE802.11a.

- IEEE802.11n:
  The IEEE802.11n Task Group was created in 2003 with the target of defining a new PHY (and the necessary MAC modifications) capable of throughput much higher than 100 Mbps. The MIMO technology, combined with OFDM, was selected for this purpose, both in the 2.4 and 5 GHz ISM bands. A variety of multi-antenna configuration and MCS are foreseen, resulting in bit rates from 6.5 to 289 Mbps in 20 MHz channels. 802.11n also allows using 40 MHz channels with bit rates as high as 600 Mbps. Although the major concepts and technical parameters of the IEEE802.11n are already fixed and known, this very complex amendment is, at the time of writing, still being drafted. It is expected that the IEEE802.11n amendment (IEEE802.11n 2007) will be accepted by mid-2009.

- IEEE802.11-VHT:
  The VHT Study Group was established in May 2007 with the target of assessing the technologies and frequency bands suitable for further extending the capabilities of WLANs in the multi-Gbps range. It is anticipated that the Study Group will propose the creation of a Task Group (the group effectively in charge of drafting the standard) by mid-2008. The technical choices have not yet been made but it is likely that OFDM and MIMO will be at the heart of the technology.

The rest of this section is devoted to analyzing the requirements, performance and challenges of the IEEE802.11n.

Objectives

The objective of the IEEE802.11n Task Group was to develop a new wireless access technology for WLAN systems that follows the evolution of the wireless applications by the end of the decade and beyond. A severe constraint was to use the existing spectrum of the ISM bands at 2.4 and 5 GHz. A direct consequence for this was that the requirements on spectral efficiency was rather demanding. As a comparison, IEEE802.11a/g delivers a peak bit rate of 54 Mbps in a 20 MHz bandwidth, which is close to 3 bits/s/Hz. Hence, it was expected that the proposed PHY technique for IEEE802.11n has a spectral efficiency on the order of 10–15 bits/s/Hz. This motivated the introduction of MIMO technology that, at the same time, had become mature at the academic level and had been demonstrated by several organizations (Kaiser *et al.* 2005). Because of the multi-path propagation in the WLAN scenarios, the combination of MIMO and OFDM was a rather natural choice: the MIMO processing can be carried out per sub-carrier, which involves a matrix model of small dimensionality.

The basic requirement for the IEEE802.11n PHY layer was to develop an air interface capable of 100 Mbps, but it was desired to provide much higher bit rates up to around 500 Mbps. These requirements called for a very high spectral efficiency.

High-level solutions

A number of techniques contribute to the high throughput of IEEE802.11n. Among others, we can mention:

- space-division multiplexing: up to four streams can be sent simultaneously;
- tighter use of OFDM (smaller guard band and cyclic prefix): 52 data sub-carriers are used instead of 48; cyclic prefix can be 400 ns instead of 800 ns;
- higher code rate: up to 5/6 instead of up to 3/4;
- diversity: exploits the availability of multiple transmit and/or receive antennas to improve range and reliability;
- channel bonding: two adjacent 20 MHz channels can be 'bonded', effectively more than doubling the peak rate;
- frame aggregation: a MAC technique that reduces the overhead for acknowledgements;
- reduced inter-frame spacing: shorter delay between transmissions reduces the time wasted between frames;
- 'Greenfield' mode: improves efficiency by eliminating support for legacy 802.11a/b/g devices (only in an all IEEE802.11n network).

6.1.2 System description

IEEE802.11n defines three formats of operation: *non-High Throughput* (HT), *HT mixed format* and *Greenfield*. The non-HT is a SISO mode, slightly different from IEEE802.11a/g that achieves 65 Mbps instead of 54 Mbps. The HT mixed format uses MIMO techniques but includes a SISO preamble to ensure compatibility with legacy devices (the compatibility only ensures that the HT devices will co-exist with the legacy systems). Finally, the Greenfield format is a purely MIMO format and cannot co-exist with other systems (they will generate interference to them). In the remaining part of this section, we will only describe the 'Greenfield' mode, which is used when only IEEE802.11n terminals and access points are present. Nevertheless, all the techniques described in this section are also applicable to HT mixed format modes.

EMERGING WIRELESS COMMUNICATION SYSTEMS

Air interface

The PHY transmission occurs in bursts, each burst consisting of three parts: a preamble for acquisition and channel estimation, a signal field to signal the current transmission mode and the payload itself. OFDM modulation and MIMO techniques are used for the payload. The reference transmitter block diagram is shown in Figure 6.1. In order to explain the transmitter functionality, we introduce the following notations:

- N_T = Number of transmit antennas
- N_R = Number of receive antennas
- N_{ES} = Number of encoded streams
- N_{SS} = Number of spatial streams
- N_{STS} = Number of space–time streams

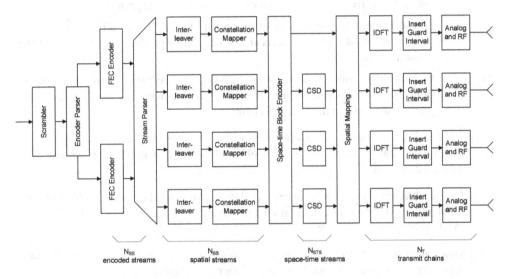

Figure 6.1 Reference transmitter block diagram for IEEE802.11n ('Greenfield' only).

The generation of the transmit signal vector is carried out by processing the transmit bit stream by means of the following sequence of blocks (Figure 6.1):

1. Scrambler: the data is scrambled to reduce the probability of long sequences of zeros or ones.

2. Encoder parser: the scrambled bits are de-multiplexed among N_{ES} encoder inputs.

3. FEC encoders: the data is encoded to enable error correction. This includes a convolutional encoder and a puncturing device.

4. Stream parser: the encoder outputs are divided into blocks that are sent to N_{SS} different interleaver and mapping devices. The sequence of the bits sent to an interleaver is called a *spatial stream*.

5. Interleaver: it interleaves the bits of each spatial stream (changes the order of the bits) to prevent long sequences of adjacent noisy bits from entering the decoder.

6. Constellation mapper: the sequence of bits in each spatial stream are mapped to constellation points in the complex plane.

7. Space–time block encoder: the constellation points from the *spatial streams* are spread into *space–time streams* using a space–time block code (note that STBC is not always used).

8. Spatial mapper: the space–time streams are mapped to transmit chains. This may include one of the following:

   – Direct mapping: the constellation points from each space–time stream are mapped directly onto the transmit chains (one-to-one mapping);
   – Spatial expansion: the vectors of constellation points from all the space–time streams are expanded via matrix multiplication to produce the input to all the transmit chains;
   – Beamforming: similarly to spatial expansion, each vector of constellation points from all the space–time streams is multiplied by a matrix of steering vectors to produce the input to the transmit chains.

9. Inverse discrete Fourier transform (IDFT): it converts a block of frequency domain constellation points to a time domain block. The IDFT is usually implemented by means of an IFFT.

10. Cyclic shift diversity (CSD) insertion: the insertion of the cyclic shifts prevents unintentional beamforming. CSD insertion may occur before or after the IDFT. Three cyclic shift types are foreseen.

11. Guard interval (GI) insertion: it pre-pends to the symbol a circular extension of itself (cyclic prefix) in order to enable frequency–domain equalization.

12. Windowing: this optionally smoothes the edges of each symbol to increase spectral decay.

Frame description

The frame format for the IEEE802.11n 'Greenfield' is illustrated in Figure 6.2. Note that the 'Greenfield' frame format is not compatible with legacy OFDM SISO systems and that the HT mixed format must be used when legacy systems coexist with IEEE802.11n systems. We will describe each part of the 'Greenfield' frame as is necessary for the topics covered in this section.

Throughout the greenfield format preamble, the same cyclic shift is applied to prevent beamforming when similar signals are transmitted on different spatial streams. For antenna j,

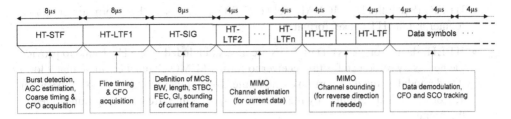

Figure 6.2 Frame of the IEEE802.11n, showing the preamble for acquisition and channel estimation, the signal field for definition of the MCS used in the subsequent payload blocks ('Greenfield' only).

the waveform of the HT-STF, HT-LTF and HT-SIG fields in the frame can be mathematically defined by:

$$s^j_{Field}(t) = \frac{1}{\sqrt{N_{STS} N^{tone}_{Field}}} \sum_{q=-Q/2}^{Q/2-1} \sum_{k=1}^{N_{STS}} \tilde{\underline{M}}^q \, [\underline{P}_{Field}]_{1\ldots k,k} \, \tilde{s}^q_k \cdot e^{j 2\pi q \Delta_F (t - T_{GI,Field} - T^k_{CS,Field})}$$

(6.1)

and the waveform of the HT-data field in the frame can be mathematically defined by:

$$s^j_{Field}(t) = \frac{1}{\sqrt{N_{STS} N^{tone}_{Field}}} \sum_{q=-Q/2}^{Q/2-1} \sum_{k=1}^{N_{STS}} \tilde{\underline{M}}^q \, \tilde{s}^q_k \cdot e^{j 2\pi q \Delta_F (t - T_{GI,Field} - T^k_{CS,Field})} \quad (6.2)$$

where the subscript *Field* takes value from one of four possible fields in the set {HT-STF, HT-LTF, HT-SIG, HT-data}, N_{STS} is the number of space–time streams, N^{tone}_{Field} is the number of tones in the given field, $\tilde{\underline{M}}^q$ is a $N_{TX} \times N_{STS}$ spatial mapping matrix per tone that is equal to the identity matrix when the channel is unknown to the transmitter, $\underline{P}$ is a 4×4 per antenna mapping matrix (only the first N_{STS} rows ad columns of $\underline{P}$ are used), $\tilde{s}$ is a vector containing the frequency domain values of the tones used in the given field, T_{GI} is the guard interval duration and T_{CS} is the cyclic shift delay. Note that, for clarity, we have not included a $rect(\cdot)$ function in Equations (6.1) and (6.2) to indicate that each field of the frame takes non-zero values only in the time interval defined in Figure 6.2. Specific values of the parameters are detailed in Tables 6.2 and 6.3.

The greenfield frame consists of the following fields:

- HT-STF (8 μs): one HT Short Training Field (HT-STF) for AGC, timing acquisition, and coarse frequency acquisition. Basically, the HT-STF transmits the same OFDM symbol $\tilde{s}$ on all spatial streams. $\tilde{s}$ has non-zero values every fourth sub-carrier and is spatially mapped on the TX antennas by $\tilde{\underline{M}}^q$, multiplied by 1 or -1 according to $[\underline{P}]_{i,k}$ and with the cyclic shift delay defined in T^k_{CS}.

- HT-LTF1 (8 μs): one double length long training field (LTF) provided as a way for the receiver to estimate the channel between each spatial mapper input and receive chain. HT-LTF1 undergoes the same processing as HT-STF.

- HT-SIG (8 µs): the HT SIGNAL Field (HT-SIG) provides all the information required to interpret the HT packet format.

- HT-LTF-data (4 or 12 µs): Data LTFs are always included in the HT frame to provide the necessary reference for the receiver to form a MIMO channel estimate that allows it to demodulate the data portion of the frame. The number of Data LTFs may be either 1, 2 or 4, and is determined by the number of space–time streams being transmitted in the frame.

- HT-LTF-extension (multiple of 4 µs): Extension LTFs provide additional reference in sounding packets so that the receiver can form an estimate of additional dimensions of the channel beyond those that are used by the data portion of the frame. This can be used in closed-loop MIMO (as in Section 2.3.2). The number of Extension LTFs may be either 0, 1, 2 or 4.

- HT-Data ($N_{symb} \cdot 4$ µs): this field contains the actual payload data.

The specific values of the parameters in Equations (6.1) and (6.2) are as follows (see also Table 6.2):

$$\underline{\underline{P}} = \begin{pmatrix} 1 & -1 & 1 & 1 \\ 1 & 1 & -1 & 1 \\ 1 & 1 & 1 & -1 \\ -1 & 1 & 1 & 1 \end{pmatrix} \qquad (6.3)$$

Table 6.2 Greenfield frame parameters.

Parameter	HT-STF	HT-LTF	HT-SIG	HT-data
N_{STS}	1	1, 2 or 4	1	1, 2, 3 or 4
N_{Field}^{tone}	12	56	56	56
$\underline{\underline{P}}_{Field}$	$\underline{\underline{P}}$	$\underline{\underline{P}}$	$\underline{\underline{P}}$	1
$\underline{s}^q$	s_{STF}	s_{LTF}	s_{SIG}	s_{Data}
$T_{GI,Field}$	16	32	16	16
$T_{CS,Field}$	see table	see table	see table	see table

Table 6.3 Cyclic shift delay for HT modes.

	T^{CS} values			
Number of space time streams	Space–time stream 1	Space–time stream 2	Space–time stream 3	Space–time stream 4
1	0 ns			
2	0 ns	−400 ns		
3	0 ns	−400 ns	−200 ns	
4	0 ns	−400 ns	−200 ns	−600 ns

EMERGING WIRELESS COMMUNICATION SYSTEMS

$$\tilde{s}_{STF} = \sqrt{1/2} \cdot [0, 0, 0, 0, 1+j, 0, 0, 0, -1-j, 0, 0, 0, 1+j,$$
$$0, 0, 0, -1-j, 0, 0, 0, -1-j, 0, 0, 0, 1+j, 0, 0, 0,$$
$$0, 0, 0, 0, -1-j, 0, 0, 0, -1-j, 0, 0, 0, 1+j,$$
$$0, 0, 0, 1+j, 0, 0, 0, 1+j, 0, 0, 0, 1+j, 0, 0, 0, 0]^T \quad (6.4)$$

$$\tilde{s}_{LTF} = [1, 1, 1, 1, -1, -1, 1, 1, -1, 1, -1, 1, 1, 1,$$
$$1, 1, 1, -1, -1, 1, 1, -1, 1, -1, 1, 1, 1, 1, 0,$$
$$1, -1, -1, 1, 1, -1, 1, -1, 1, -1, -1, -1, -1, -1,$$
$$1, 1, -1, -1, 1, -1, 1, -1, 1, 1, 1, 1, -1, -1]^T \quad (6.5)$$

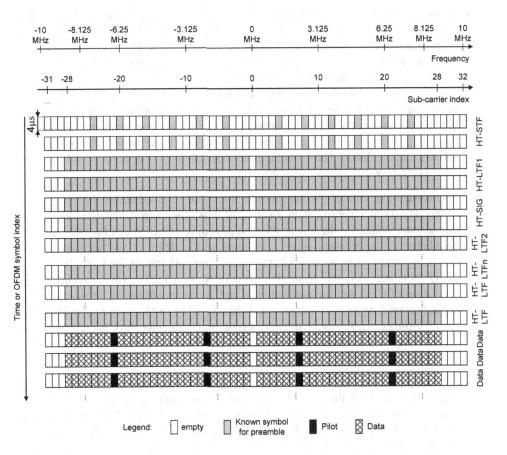

Figure 6.3 Time-frequency illustration of the IEEE802.11n frame.

6.1.3 Main challenges and usual solutions

Acquisition and channel estimation strategy

As for every standard, a successful detection and an accurate acquisition and channel estimation in IEEE802.11n are crucial to achieve good performance and high throughput. This is even more so for the HT MIMO-OFDM that can be plagued by ICI due to uncompensated CFO, SCO or IQ imbalance and inter-stream interference due to inaccuracy in the MIMO channel estimation. Note also that MIMO channel estimation implies the estimation of $N_R \times N_T$ channels as opposed to 1 channel for SISO transmission. Thanks to the structure of the IEEE802.11n frame (Figure 6.3), the acquisition and channel estimation strategy can reuse many techniques and hardware from the legacy IEEE802.11a/g SISO-OFDM devices. There are some differences, however, that are needed to enable the MIMO channel estimation. The following strategy, illustrated in Figure 6.2, can be applied successfully to demodulate the IEEE802.11n payload data:

- Coarse acquisition: on each receive antenna, the signal received during the transmission of the HT-STF is repetitive with a repetition period of 800 ns, independently of the channel, just as in IEEE802.11a/g. Hence, the same technique for the coarse acquisition of timing and CFO can be used: an auto-correlation of the received signal with itself with a delay of 16 baseband samples (800 ns) can be used to provide an estimate of timing and CFO. Note that no performance improvement should be expected from the multi-antenna transmission because there is no diversity optimization (the power is blindly split among the transmit antennas). At the receive side, some improvement on the coarse acquisition can be achieved if the acquisition is performed on several receive antenna and combined. The advantage due to multi-antenna acquisition will manifest itself as a processing gain.

- Fine acquisition: on each receive antenna, the signal received during the transmission of the HT-LTF is repetitive with a period of 4 μs, just as in IEEE802.11a/g. Hence, the same technique for the fine acquisition of timing and CFO can be used. An auto-correlation of the received signal with itself with a delay of 64 baseband samples (4 μs), as described in Section 5.5, provides a fine estimate of the CFO. In the case of IQ imbalance, the technique described in Section 5.9 to provide accurate joint estimation of CFO and the parameters α and β can be applied. A cross-correlation (as in Section 5.6) with the transmitted field (HT-LTF) can be used to provide a finer timing estimate. This must be done with care, however, since the CSD results in time offsets between the signals transmitted by the different TX antennas and the correct peak must be selected.

- Channel estimation: once correct timing, CFO and IQ parameters have been acquired and compensated for, the MIMO channel estimation can take place. It is important to realize that no channel estimation can be made based on HT-LTF1 alone if more than one transmit antenna is used. The HT-LTF2 is needed if two TX antennas are used and HT-LTF3 and HT-LTF4 are needed if 3 or 4 TX antennas are used. This will be explained in detail in the next sub-section.

The details of the coarse and fine acquisition and channel estimation is provided in the next sub-sections.

EMERGING WIRELESS COMMUNICATION SYSTEMS

Details of the coarse timing and CFO acquisition

On each receive antenna, the signal received during the transmission of the HT-STF is repetitive with a repetition period of 800 ns, independently of the channel, as long as the channel impulse response is shorter than 800 ns. Hence, an auto-correlation of the received signal with itself with a delay of 16 baseband samples (800 ns), as described in Section 5.4, can be used to provide an estimate of timing and CFO. The unambiguous CFO acquisition range is $]-1/2 \cdot 800 \text{ ns}, 1/2 \cdot 800 \text{ ns}[=]-625 \text{ kHz}, 625 \text{ kHz}[$. If only one receive antenna is used, the method of Section 5.4 directly applies. With multiple receive antennas, several ways of exploiting them can be conceived.

Coarse timing acquisition

Many ad-hoc strategies can be used to estimate the peak of the auto-correlation in a MIMO set-up. The auto-correlation on each receive antenna branch can be expressed as follows:

$$AC_i[n] = \sum_{m=0}^{M-1} r_i^*[n+m-L_{STF}]r_i[n+m] \qquad (6.6)$$

where $r_i[n]$ is the received preamble sequence on the receive antenna i corrupted by the multi-path channel and AWGN, L_{STF} is the duration of the elementary sequence in the HT-STF and M is matched to the length of HT-STF, taking into account that part of the HT-STF has been consumed for AGC. The following are the most representative methods of acquiring timing information from multiple receive antennas:

- post-averaging: the time indices of the peak of the auto-correlation on each receive antenna are averaged. This can be expressed as:

$$\hat{n}_{AC} = \frac{1}{N_T} \sum_{i=1}^{N_T} \arg\max_n (AC_i[n]) \qquad (6.7)$$

- pre-averaging: the auto-correlations on each receive antenna are first averaged, resulting in a unique auto-correlation in which the location of the peak provides the timing estimation:

$$\hat{n}_{AC} = \arg\max_n \left(\frac{1}{N_T} \sum_{i=1}^{n_T} (AC_i[n]) \right) \qquad (6.8)$$

- minimum time index: the minimum of the time indices of the peak of the auto-correlation on each receive antenna provides the timing estimation:

$$\hat{n}_{AC} = \min_i (\arg\max_n (AC_i[n])) \qquad (6.9)$$

- time index of the maximum peak: the peaks are searched on each receive antenna. The peak corresponding to the maximum auto-correlation amplitude is selected and its time index provides the timing estimation:

$$n_i = \arg\max_n (AC_i[n]) \qquad (6.10)$$

$$\hat{n}_{AC} = n_{\arg\max_i (AC_i[n_i])} \qquad (6.11)$$

The performance of these four coarse timing acquisition methods is compared in Figure 6.4. The pre- and post-averaging methods provide at the same time the best performance and an unbiased behavior at all SNRs. Pre-averaging has a slightly better behavior. The 'minimum time index' and 'time index of the maximum peak' methods do not perform as well: their histogram is more spread. The histogram for the single antenna receiver is also shown in Figure 6.5 for comparison. The dramatic improvement resulting from combining the auto-correlation from several antennas is clearly visible (in Figure 6.5, the histogram is less peaky and the maximum value is lower).

Coarse CFO acquisition

The coarse CFO can be estimated from the HT-STF field by measuring the angle of the accumulated auto-correlation at the estimated timing instant, as described in Section 5.5. The correct way to combine the different receive antennas is first to sum the auto-correlation value at the optimum timing $\hat{n}_{AC}$ and then compute the CFO by means of Equation (5.15). In this way, the relative amplitudes of the auto-correlation serve as 'weights' in the averaging process.

The performance of the auto-correlation-based CFO estimation on the HT-STF is illustrated in Figure 6.6 where the variance of the CFO estimation error is plotted as a function of the SNR. It can be observed that the variance gets smaller as the number of receive antennas increases, which can be interpreted as a processing gain. Note that the little flooring effect is attributable to the fact that the timing point used in the CFO estimation is the estimated timing point. When the ideal timing point is used, the flooring disappears, as can be seen from the dashed curves.

Details of fine timing and CFO estimation

Fine timing and CFO acquisition

The fine estimation of timing and CFO is done on the HT-LTF field. This is a straightforward multi-antenna extension of the auto-correlation and cross-correlation methods described in Sections 5.6 and 5.7, taking into account the issues discussed in the previous sub-sections about coarse timing and coarse CFO estimation.

Joint estimation of CFO and IQ Imbalance

If IQ imbalance is present in the MIMO receiver, it must be estimated and compensated before the MIMO channel estimation. During the reception of the HT-LTF1 field, each receive antenna can independently apply a joint CFO and IQ imbalance parameters estimation as if they were independent SISO antennas. The technique described in Section 5.9 can be applied.

Details of the MIMO channel estimation

This sub-section deals with the MIMO channel estimation when the IEEE802.11n preamble is transmitted. An efficient channel estimation method is to estimate the channel response in the time domain since it is known that the impulse response has a limited duration. It follows that the number of coefficients to estimate (L) is much smaller than the number of used

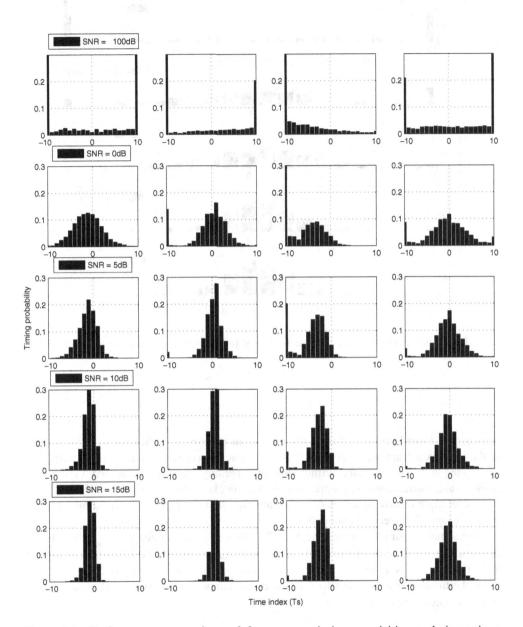

Figure 6.4 Performance comparison of four coarse timing acquisition techniques in a IEEE802.11n MIMO set-up with one transmit antenna and four receive antennas (from top to bottom: SNR = −100, 0, 5, 10 and 15 dB; from left to right: post-averaging, pre-averaging, minimum time index and time index of the maximum peak).

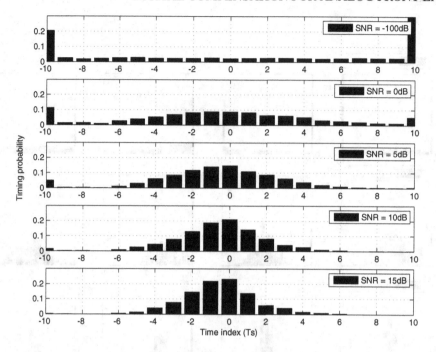

Figure 6.5 Performance of coarse timing acquisition with one receive antenna.

sub-carriers. In addition to reducing the number of unknown coefficients to estimate, this method has the advantage of smoothing the frequency response, which is a consequence of windowing in the time domain. Without loss of generality, we will assume that N_T transmit antennas and one receive antenna are used. We further exploit the constraint that $L \ll Q$. The case of multiple receive antennas can straightforwardly be extended by repeating the same processing on all the receive antennas.

The signal transmitted during the HT-LTF fields is characterized by $\tilde{\underline{s}}_{k,u}$ which is the frequency domain vector of the uth HT-LTF field transmitted on the kth antenna as given by Equation (6.5). Note that $\tilde{\underline{s}}_{k,u}$ is made of 56 non-zero values out of 64. The goal of the channel estimation is to estimate the N_T time domain impulse responses $\underline{h}_k$ which are related to their frequency responses $\tilde{\underline{h}}_k$ by

$$\tilde{\underline{h}}_k = \bar{\underline{\underline{F}}} \cdot \underline{h}_k \quad (k = 1, \ldots, N_T) \tag{6.12}$$

where $\bar{\underline{\underline{F}}}$ is a made of the L first columns of a normalized Fourier matrix $\underline{\underline{F}}_Q$ with elements $[\underline{\underline{F}}_Q]_{ab} = e^{-j2\pi ab/N}/\sqrt{Q}$. Note that it is precisely this reduction in the number of columns (and rank) of the Fourier matrix that fixes the length of the estimated impulse response.

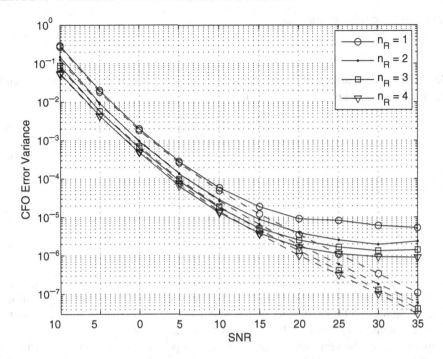

Figure 6.6 CFO estimation error variance from auto-correlation on the received HT-STF, for up to four antennas (solid lines: with estimated timing instant; dashed lines: ideal timing instant).

The input–output model for the channel estimation can then be expressed as:

$$\underline{\tilde{y}}_u = \sum_{k=1}^{N_T} \underline{\underline{\tilde{\Lambda}}}_{k,u} \cdot \underline{\tilde{h}}_k + \underline{\tilde{w}} \qquad (6.13)$$

$$= \sum_{k=1}^{N_T} \underline{\underline{\tilde{\Lambda}}}_{k,u} \cdot \underline{\underline{\bar{F}}} \cdot \underline{h}_k + \underline{\tilde{w}} \qquad (6.14)$$

where $\underline{\tilde{w}}$ is a vector of AWGN noise in the frequency domain and:

$$\underline{\underline{\tilde{\Lambda}}}_{k,u} = \operatorname{diag}(\underline{\tilde{s}}_{k,u}) \qquad (6.15)$$

To derive the ML solution, it is convenient to express the model in matrix form as follows. By defining:

$$\underline{\underline{\tilde{\Lambda}}}_u := \left[\underline{\underline{\tilde{\Lambda}}}_{1,u} \cdots \underline{\underline{\tilde{\Lambda}}}_{N_T,u} \right] \qquad (6.16)$$

$$\underline{\underline{F}} := \underline{\underline{I}}_{N_T \times N_T} \otimes \underline{\underline{\bar{F}}} \qquad (6.17)$$

$$\underline{h} := [\underline{h}_1^T \cdots \underline{h}_{N_T}^T]^T \qquad (6.18)$$

Equation (6.13) can be rewritten as:

$$\underline{\tilde{y}}_u = \underline{\underline{\tilde{\Lambda}}}_u \cdot \underline{\underline{F}} \cdot \underline{h} + \underline{\tilde{w}} \qquad (6.19)$$

Under the condition of AWGN noise, the ML solution of this linear problem is then the least-squares solution given by (see Appendix B):

$$\underline{\hat{h}} = \underline{\underline{B}} \cdot \underline{\tilde{y}}_u \qquad (6.20)$$

$$\underline{\hat{\tilde{h}}} = \underline{\underline{F}} \cdot \underline{\hat{h}} \qquad (6.21)$$

where matrix $\underline{\underline{B}}$ is obtained by:

$$\underline{\underline{A}} = \underline{\underline{F}}^H \cdot (\underline{\underline{\tilde{\Lambda}}}_u)^H \cdot \underline{\underline{\tilde{\Lambda}}}_u \cdot \underline{\underline{F}} \qquad (6.22)$$

$$\underline{\underline{B}} = (\underline{\underline{A}})^{-1} \cdot \underline{\underline{F}}^H \cdot (\underline{\underline{\tilde{\Lambda}}}_u)^H \qquad (6.23)$$

In order to estimate $\underline{\hat{h}}$, which has dimension LN_T, the rank of $\underline{\underline{A}}$ must be higher than or equal to $L \cdot N_T$. This requirement indicates that the maximum channel length that can be estimated is equal to 56 divided by the number of TX antennas (56, 28, 18 and 14 for one, two, three and four antennas, respectively) (Horlin and Van der Perre 2004).

However, since the same set of frequency domain pilot data is sent on all antennas, the rank of $\underline{\underline{A}}$ can be affected. Indeed, a simple inspection of Equation (6.22) shows that if all the $\underline{\underline{\tilde{\Lambda}}}_{k,u}$ were equal, the rank of $\underline{\underline{A}}$ degenerates to L and the multi-channel estimation of N_T channels of length L is not possible. In the HT-LTF, the $\underline{\underline{\tilde{\Lambda}}}_{k,u}$ are not exactly equal. They are actually derived from the HT-LTF vector $\underline{\tilde{s}}$ by:

$$\underline{\underline{\tilde{\Lambda}}}_{k,u} = p_{k,u} \text{diag}(\underline{\tilde{c}}_k) \cdot \text{diag}(\underline{\tilde{s}}_{k,u}) \qquad (6.24)$$

where $\underline{\tilde{c}}_k$ is an antenna dependent CSD and $p_{k,u}$ is a coefficient from matrix $\underline{\underline{P}}$, equal to 1 or -1, which is both dependent on the antenna k and the index u of the HT-LTF field ($u = 1, \ldots, 4$). It is crucial to understand the effect of the CSD (Equation (6.24)) on the rank of $\underline{\underline{A}}$ ($p_{k,u}$ is a scalar and, hence, does not impact the rank of $\underline{\underline{A}}$). For a single transmit antenna, the rank of $\underline{\underline{A}}$ is L, which is sufficient. When antennas are added, transmitting the same symbol with CSD, the only difference between any two $\underline{\underline{\tilde{\Lambda}}}_{k,u}$ is due to the CSD. Equation (6.13) indicates that the CSD will increase the rank of $\underline{\underline{A}}$ only by the value of the maximum of the sample shifts caused by the CSD. Since the CSD takes on the values from $\{0, -400$ ns, -200 ns, -600 ns$\}$ (which amounts to $\{0,8,4,12\}$ samples, respectively), the rank of $\underline{\underline{A}}$ is constrained by the CSD for typical values of L. An illustration of this is given in Figure 6.7, which shows the rank of $\underline{\underline{A}}$ for the case of two transmit antennas and the CSD applied to the second antenna varies from 0 to $Q/2$. It can be observed that, for values of the CSD smaller than L, the rank of $\underline{\underline{A}}$ does not reach its full value of $2L$ which is needed to estimate two channels of length L.

A solution to this problem is to exploit the repetition of the preamble sequence over several HT-LTF fields, with the matrix $\underline{\underline{P}}$ playing the role of a spreading matrix, with the chip period extending over the duration of the field. The signal on which to perform channel

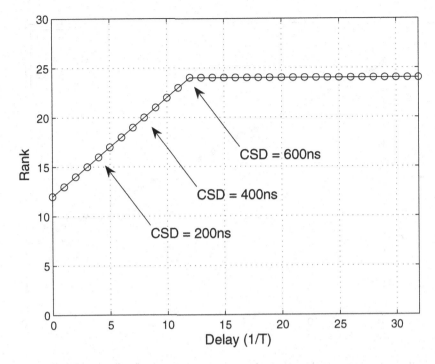

Figure 6.7 Effect of the Cyclic Shift Delay on the rank of $\underline{\underline{A}}$ in the ML solution for channel estimation ($N_T =$ two transmit antennas and channel length $L = 12$). The CSD is varied from 0 to $Q/2 = 32$. For CSD smaller than L, the rank is not sufficient to estimate two channels of length L. $1/T$ is the baseband sample rate.

estimation can then be expressed as:

$$\tilde{\underline{z}}_k = \frac{1}{U} \sum_{u=1}^{U} p_{k,u} \tilde{\underline{y}}_u \qquad (6.25)$$

where U is the number of HT-LTF. The channel estimate from transmit antenna k can be found by applying the ML solution derived above (Equation (6.20)) to $\underline{z}_k$. This procedure transforms the multi-channel estimation in N_T channel estimations. Hence, it must be applied separately for all transmit antennas. The performance of the MIMO channel estimation is illustrated in Figure 6.8. The number of transmit antennas for this simulation is one, two or four. The MSE is approximately the same for these three cases because the loss in transmit power per channel is compensated by the increase in number of time slots used for the channel estimation. No CFO was added in these simulations.

Estimation of residual CFO/SCO for tracking

In Section 5.11, we analyzed the tracking loop for a SISO OFDM system. The tracking loop for SISO-OFDM is based on a measure of the rotation of pilots symbols that are multiplexed

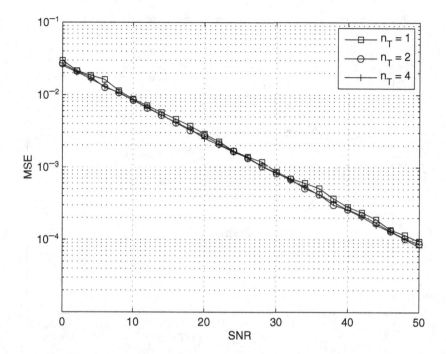

Figure 6.8 Performance of the ML MIMO channel estimator for IEEE802.11n for one, two and four transmit antennas. The MSE remains approximately constant when more transmit antennas are used because the number of time slots, hence the received energy, increases with the number of transmit antennas.

in the frequency domain with the data symbols. The same concept is used in IEEE802.11n but with the addition of the spatial dimension. Therefore, the estimators that were derived in Equations (5.108) and (5.109) must be extended to take into account the fact that pilots are inserted in every stream. The resulting ML CFO estimator for a MIMO system is then found to be:

$$\widehat{\Delta \omega}[n] = \frac{1}{2\pi Q} \angle \left(\sum_{p=1}^{N_P} \{ (\underline{\tilde{y}}^{q_p}[n])^H \cdot \underline{\underline{\tilde{H}}}^{q_p}[n] \cdot \underline{\tilde{p}}^{q_p}[n] \} \right) \quad (6.26)$$

where q_p is the pilot index, $\underline{\tilde{p}}^{q_p}$ is the vector of pilots transmitted on all spatial streams at sub-carrier q_p, $\underline{\tilde{y}}^{q_p}$ is the vector of signals received on all receive antennas at sub-carrier q_p and $\underline{\underline{\tilde{H}}}^{q_p}$ is the MIMO channel matrix at sub-carrier q_p. Note that this is not the same as the propagation channel matrix since the pilots are inserted in the data-field at the transmitter before the pre-coding and CSD. Just as for SISO-OFDM, the effect of CFO in the frequency domain is to rotate the phases of all sub-carriers on all streams by the same amount (ICI is also generated). The spatially extended SCO estimator reads:

$$\hat{\delta} = \frac{1}{\pi N_P (N_P - 1)} \sum_{p_1=1}^{N_P} \sum_{p_2=1}^{N_P} \frac{\angle (\tilde{e}^{q_{p_1}}[n](\tilde{e}^{k_{p_2}}[n])^*)}{k_{p_1} - k_{p_2}} \quad (6.27)$$

EMERGING WIRELESS COMMUNICATION SYSTEMS

where:

$$\tilde{e}^q[n] = (\tilde{\underline{y}}^q[n])^* \cdot \tilde{\underline{\underline{H}}}^q[n] \cdot \tilde{\underline{p}}^q[n] \qquad (6.28)$$

This estimator actually averages all phase differences between all possible pairs of pilots. This can be shown to be equivalent to finding the slope of the phase shifts across the subcarriers by linear regression. The CFO and SCO estimators (6.26) and (6.27) can then feed CFO or SCO tracking loops similar to the one presented in Sections 5.11.2 and 5.11.2.

It is important to mention that these tracking loops are actually made up of MIMO estimators (6.26) and (6.27) but that the compensation part is a SISO compensation (a single phase error and sampling clock is tracked). This is valid as long as the same LO(s) and sampling clocks are used in all the receive antenna branches. It also relies on the sharing of LOs and sampling clocks at the transmitter side.[3]

6.1.4 Compensation of non-reciprocity

Section 4.7.3 introduced the degradation caused by non-reciprocity on certain MIMO schemes relying on reciprocity between the reverse and forward links for channel estimation. Since it is not economical to design and manufacture wireless terminals that are perfectly reciprocal, practical systems may use on-line calibration techniques to measure the non-reciprocity and compensate it digitally. The measurement part of this procedure is technically difficult and a possible scheme will be presented here. The digital compensation part is trivial (usually a multiplication with a complex coefficient) and will not be addressed here.

A generic approach to estimate the non-reciprocity offsets is to use an existing (or an additional) transceiver as a signal source to measure other receivers and as a receiver to measure other transmitters. In Kaiser *et al.* (2005, Chapter 32), a method with an auxiliary transceiver is used to measure a multi-antenna front-end. Liu *et al.* (2006) proposed a method whereby no additional front-end is needed to measure the MIMO transceiver. We will describe this method here for a MIMO transceiver with three antennas; the technique is easily extended to an arbitrary number of antennas.

Figure 6.9 shows the block diagram of a MIMO transceiver with calibration hardware. In addition to the three transceivers, there is a network of switches and an attenuator. This network allows the first transceiver to be connected directly to the others and to serve as a signal source or signal analyzer. The steps needed for the calibration are the following:

- Step 1: $TX_1 \Rightarrow RX_2$, which provides measurement of TF_{12};

- Step 2: $TX_1 \Rightarrow RX_3$, which provides measurement of TF_{13};

- Step 3: $TX_2 \Rightarrow RX_1$, which provides measurement of TF_{21};

- Step 4: $TX_3 \Rightarrow RX_1$, which provides measurement of TF_{31}.

---

[3] Interestingly, in the uplink of an SDMA system, the MIMO tracking loop as we just described could not be applied since the transmitters (different users) cannot have the same LOs and sampling clocks. Separate tracking loops must then be used for each received stream.

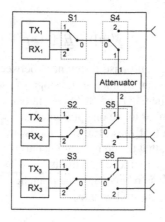

Figure 6.9 MIMO transceiver with reciprocity calibration circuit (switches show transceiver 1 as 'source').

The four transfer functions just described contain also the effect of the switches and the attenuator as follows:

$$TF_{12} \approx d_{TX_1} \cdot S4_{01} \cdot A_{12} \cdot S5_{10} \cdot d_{RX_2} \tag{6.29}$$

$$TF_{13} \approx d_{TX_1} \cdot S4_{01} \cdot A_{12} \cdot S6_{10} \cdot d_{RX_3} \tag{6.30}$$

$$TF_{21} \approx d_{TX_2} \cdot S5_{01} \cdot A_{21} \cdot S4_{10} \cdot d_{RX_1} \tag{6.31}$$

$$TF_{13} \approx d_{TX_3} \cdot S6_{01} \cdot A_{21} \cdot S4_{10} \cdot d_{RX_1} \tag{6.32}$$

where we use the following notations:

- TF_{ij} represents the transfer function from transmitter i to receiver j;
- d_{TX_i} represents the transfer function of transmitter i;
- d_{RX_i} represents the transfer function of receiver i;
- Sk_{ij} represents the transfer function of switch k, from connector i to j;
- A_{ij} represents the transfer function of the attenuator, from connector i to j.

From these relations we derive by simple division the calibration factors:

$$C_1 := 1, \tag{6.33}$$

$$C_2 := \frac{TF_{12}}{TF_{21}} = \frac{S4_{01} \cdot A_{12} \cdot S5_{10}}{S5_{01} \cdot A_{21} \cdot S4_{10}} \cdot \frac{d_{TX_1} d_{RX_2}}{d_{TX_2} d_{RX_1}} \tag{6.34}$$

$$C_3 := \frac{TF_{13}}{TF_{31}} = \frac{S4_{01} \cdot A_{12} \cdot S6_{10}}{S6_{01} \cdot A_{21} \cdot S4_{10}} \cdot \frac{d_{TX_1} d_{RX_3}}{d_{TX_3} d_{RX_1}} \tag{6.35}$$

What we actually need are the calibration coefficients:

$$C_2 = \frac{d_{TX_1} d_{RX_2}}{d_{TX_2} d_{RX_1}} \tag{6.36}$$

EMERGING WIRELESS COMMUNICATION SYSTEMS

and
$$C_3 = \frac{d_{TX_1} d_{RX_3}}{d_{TX_3} d_{RX_1}} \quad (6.37)$$

so that we have a requirement on the calibration hardware:

$$\frac{S4_{01} \cdot A_{12} \cdot S5_{10}}{S5_{01} \cdot A_{21} \cdot S4_{10}} \approx 1 \quad (6.38)$$

$$\frac{S4_{01} \cdot A_{12} \cdot S6_{10}}{S6_{01} \cdot A_{21} \cdot S4_{10}} \approx 1 \quad (6.39)$$

which is equivalent to requiring that the switches and attenuators in Equation (6.38) are reciprocal; this is much easier to achieve than to have the complete transceiver reciprocal. At this point, it is worth mentioning that the transfer function of devices, including switches and attenuators are reciprocal only when the impedances do not change according to the direction of the signal. Therefore, the requirements in Equation (6.38) are not automatically satisfied.

It is easy to show that, having coefficients C_2 and C_3, we can compensate for the non-reciprocity of the three transceivers. Indeed, if we multiply the transmitted signals of antenna 1, 2 and 3 by C_1, C_2 and C_3, respectively, and then compute the ratio of the equivalent transfer function of the transmitter to the transfer function of the receiver, we obtain:

$$R_1 = \frac{d_{TX_1}}{d_{RX_1}} \cdot C_1 = \frac{d_{TX_1}}{d_{RX_1}} \quad (6.40)$$

$$R_2 = \frac{d_{TX_2}}{d_{RX_2}} \cdot C_2 = \frac{d_{TX_2}}{d_{RX_2}} \cdot \frac{d_{TX_1} d_{RX_2}}{d_{TX_2} d_{RX_1}} = \frac{d_{TX_1}}{d_{RX_1}} \quad (6.41)$$

$$R_3 = \frac{d_{TX_3}}{d_{RX_3}} \cdot C_3 = \frac{d_{TX_3}}{d_{RX_3}} \cdot \frac{d_{TX_1} d_{RX_3}}{d_{TX_3} d_{RX_1}} = \frac{d_{TX_1}}{d_{RX_1}} \quad (6.42)$$

$$(6.43)$$

We see that all antennas have the same ratio between transmit and receive transfer functions, which is indeed equivalent to having matched transceivers, free of non-reciprocity.

6.2 3GPP Long-term evolution

6.2.1 Context

History

Wideband CDMA has been initially selected as the core multi-access technology for 3G wireless cellular systems because it offers a high system capacity and supports interesting networking properties such as soft hand-over (3GPP Release 99). The system evolution towards a packet access has been made to ensure the competitiveness in the mid-term future (3GPP Release 6, HSDPA/DSUPA).

The Evolved-UTRA or 3GPP long-term evolution (LTE) is a major step undertaken by the 3GPP standardization committee to meet the user expectations in a 10 years perspective and beyond. The 3GPP work on the evolution of the 3G mobile systems started with a workshop organized in November 2004 on the envisioned radio access network. Wideband CDMA has been replaced by the OFDMA multi-access technology that can deal better with the time dispersive channels. A feasibility study, performed until December 2006, confirmed that the selected technologies can satisfy the defined objectives. The system specifications are currently being fixed. 3GPP LTE is a candidate for the worldwide IMT-Advanced standardization initiative defined to expand IMT 2000.

Objectives

The objective of the 3GPP LTE is to provide a wireless access competitive to existing fixed line accesses, at substantially reduced user/operator cost compared to current radio access technologies. Assuming the convergence toward the use of the Internet Protocol, the enhancements should mainly be developed for packet-based services.

A list of minimal requirements has been set up to guarantee the competitiveness of the new system.

- High data rates:

  peak data rates up to 100 Mbps in the downlink and up to 50 Mbps in the uplink should be supported.

- Improved system capacity:

  the average user throughput should be three to four times higher in the downlink and two to three times higher in the uplink compared to the previous Release 6 (HSDPA/HSUPA). At the edge of the cell, the throughput should be two to three times higher.

- Reduced latency:

  the user plane round-trip time should be smaller than 10 ms and the channel set-up delay should be smaller than 100 ms.

- High mobility:

  the system should work optimally, i.e. at the highest data rates, for low terminal speeds from 0 to 15 km/h, keep a high performance for higher terminal speeds between 15 and 120 km/h, and still maintain the communication at terminal speeds from 120 km/h to 350 km/h.

- High coverage:

  the throughput, spectrum efficiency and mobility targets defined above should be met for 5 km cells, and with a slight degradation for 30 km cells. Cells range up to 100 km should not be precluded.

- Co-existence and inter-working (hand-over) with the 3GPP Release 6 radio access technology.

6.2.2 System description

Air interface

Compared to the 3GPP Release 6 based on wideband CDMA, the physical layer of the Evolved UTRA is based on OFDM mainly because it offers a higher spectral efficiency, it supports a flexible allocation of the spectrum, and it is easily combined with MIMO processing. Figure 6.10 illustrates how blocks of transmitted symbols are modulated.

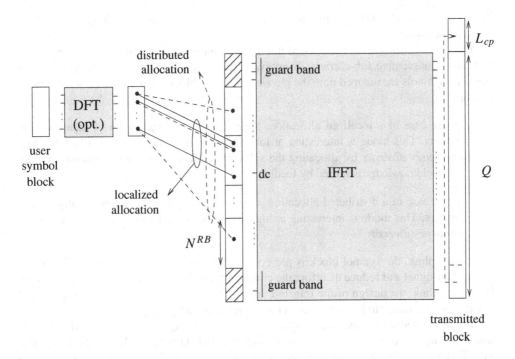

Figure 6.10 3GPP LTE air interface.

The communication bandwidth can vary from 1.25 MHz to 20 MHz, making the system scalable as the transmit/receive computational complexity can be adjusted to the capacity needs. The sub-carrier spacing is fixed to 15 kHz for all values of the communication bandwidth. Consequently the number of sub-carriers Q varies proportionally with the bandwidth (from 128 to 2048). This choice offers two advantages:

1. the robustness against the mobility is kept, whatever the communication bandwidth (the symbol block duration is constant);

2. the implementation of dual mode Release 6/Evolved-UTRA terminals is facilitated (a single clock frequency is necessary).

The DC carrier is conventionally placed in the middle of the block. Large guard bands are placed at both sides of the spectrum to respect the bandwidth requirements (approximately

60% of the sub-carriers are occupied). The cyclic prefix length L_{cp} can take two values to better adapt the necessary overhead to the propagation channel conditions:

- first, a short cyclic prefix of 4.7 μs can be selected to reduce the overhead in the typical channel delay spread cases;

- second, a long cyclic prefix of 16.7 μs can be selected to cope with the high channel delay spreads possible in the case of very large cells (up 100 km radius cells).

The overall block duration is equal to 71 μs or to 83 μs in the short and long cyclic prefix cases, respectively.

The block of sub-carriers is divided into a set of physical resource blocks composed of $N^{RB} = 12$ consecutive sub-carriers, or equivalently of 128 kHz spectrum occupation. The user data symbols are mapped onto the physical resource blocks in a localized or distributed fashion.

- In the case of a localized allocation, the users are allocated a set of adjacent sub-carriers. This mode is interesting at low-mobility since the system can benefit from *multi-user diversity* by allocating the sub-carrier blocks optimally based on a partial channel knowledge acquired by feedback.

- In the case of a distributed allocation, the users are allocated a set of equally spaced carriers. This mode is interesting at high mobility since the system can benefit from *frequency diversity*.

In the uplink, the symbol block is pre-coded with a DFT to decrease the PAPR of the transmitted signal and reduce therefore the constraints on the design of the analog front-end. In the downlink, the design of the transmit analog front-end is less critical and the symbol block is not pre-coded to keep the receiver at a minimum complexity.

The supported downlink data modulation schemes are QPSK, 16QAM and 64QAM, while the supported uplink data modulation schemes are QPSK, 8PSK and 16QAM. The use of MIMO techniques is encouraged, with possibly up to four antennas at the mobile terminal, and four antennas at the base station. Cyclic-delay diversity, SDM with and without codebook-based pre-coding techniques are specified in the standard. The turbo-code specified in the previous system releases is re-used in order not to waste the existing expertise.

Frame description

Figure 6.11 illustrates how the radio frame is defined in the 3GPP LTE system. In this description, we focus on the frame structure of type 1 dedicated to the TDD mode.

Each radio frame is 10 ms long and consists of 20 slots of 0.5 ms duration. A sub-frame, defined as two consecutive slots, is either allocated for downlink or uplink transmission. The sub-frames 0 and 5 are mandatorily allocated for downlink transmission. Therefore, the frame is divided in two main parts, each composed of a downlink and uplink transmission sub-part. For TDD operation, the last downlink OFDM block(s) in a sub-frame immediately preceding a downlink-to-uplink switch point can be reserved for guard time and consequently not transmitted (up to 12 OFDM blocks in the case of a short cyclic prefix and up to 10 OFDM blocks in the case of a long cyclic prefix can be reserved for guard time). Thanks to

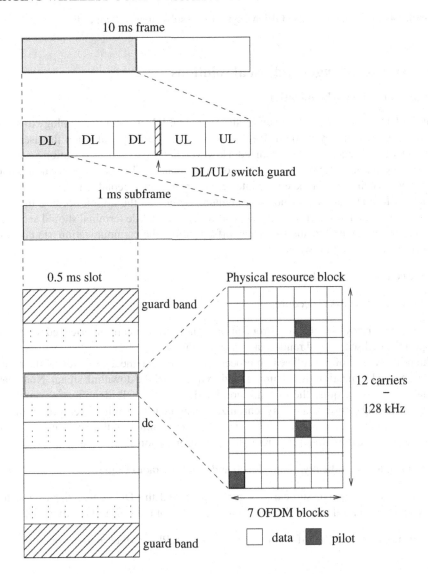

Figure 6.11 3GPP LTE frame (type 1, TDD mode, short cyclic prefix, 1 antenna port).

the guard time, the mobile terminals can switch from receive to transmit mode and the time synchronization between terminals is made possible.

Each slot is described by a resource grid in the frequency and time domains. The number of OFDM symbols per slot is either seven (short cyclic prefix) or six (long cyclic prefix). A physical resource block is composed 12 consecutive sub-carriers in the frequency domain (128 kHz bandwidth) and spans over one slot in the time domain. In case of a short cyclic prefix, two pilot symbols are placed on the OFDM block 1 (sub-carriers 1 and 7) and two pilot symbols are placed on the OFDM symbol 5 (sub-carriers 4 and 10) of each resource block.

6.2.3 Main challenges and usual solutions

Multi-user uplink synchronization

The 3GPP LTE communication system relies on the OFDMA access technology to separate the users in the frequency domain. Ideally the system is orthogonal when the user signals are perfectly time/frequency synchronized. However, any small synchronization error causes inter-user interference and results in a loss of performance. Therefore, the terminals should be finely pre-synchronized to keep the interference below an acceptable level.

The synchronization is a particularly challenging task in the uplink because the mobile terminals are located at different varying positions. Each mobile terminal should acquire and compensate its relative time and frequency offsets before the communication can take place. This is usually achieved in two steps:

- the cell search procedure;

- the random access procedure.

Figure 6.12 describes how the synchronization signals are typically interleaved in the frame to support the cell search and random access procedures.

The primary goal of the cell search procedure is to enable the acquisition of the received timing (the synchronization signal timing) and frequency of the downlink signal. Note that the terminals must also acquire the transmission bandwidth, the cell identification number, the radio frame timing (more than one synchronization signal is transmitted per radio frame), the cyclic prefix length .... Synchronization signals are transmitted in the downlink to facilitate the cell search. The downlink time/frequency synchronization signals are:

- transmitted periodically in the slots 0 and 10 of the radio frame;

- located on 72 active sub-carriers centered around the DC sub-carrier (the terminal detects the central part of the spectrum regardless of the transmit bandwidth);

- generated from a Zadoff–Chu sequence (low PAPR, perfect auto-correlation properties).

The time synchronization is usually performed by cross-correlating the received signal with the expected transmitted preamble and by detecting the peak corresponding to the preamble time-of-arrival. The cross-correlator is the optimal ML estimator when the transmitted preamble if known and only corrupted by additive white Gaussian noise (no multi-path) (Meyr et al. 1998). It benefits from the perfect auto-correlation properties of the Zadoff–Chu sequences. The performance of the time estimator is, however, strongly limited by the presence of CFO. The time and CFO can be estimated jointly by cross-correlating the received signal with multiple versions of the preamble, each affected by a different value of the CFO, and by selecting the preamble giving rise to the highest peak. The time/CFO estimate can be performed over multiple preambles to improve the accuracy of the estimate and/or share the complexity over the time.

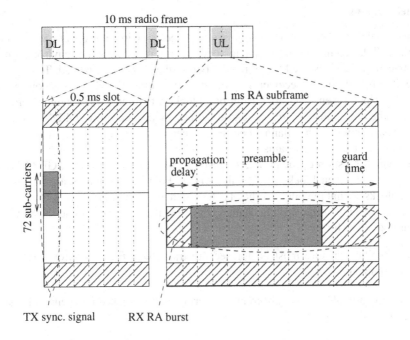

Figure 6.12 3GPP LTE time/frequency synchronization. The dark blocks represent the synchronization signals.

The goal of the random access procedure is to synchronize the transmissions of the different terminals in the time (the remaining asynchronism must be included in the cyclic prefix). Random access bursts are transmitted in the uplink to enable the time synchronization. The uplink random access bursts are:

- composed of a cyclic prefix, a preamble and a guard time (included to cope with the time uncertainty);
- sent during the duration of one sub-frame;
- located on 72 adjacent sub-carriers (six resource blocks) separated from the other terminal random access bursts in the frequency domain;
- generated from a Zadoff–Chu sequence (low PAPR, perfect auto-correlation properties).

When receiving a random access burst, the base station determines if the terminal needs a transmit time adjustment and communicates it to the terminal together with the terminal resource allocation through the downlink control channel. Note that the user terminal CFO is pre-compensated in the uplink based on the estimation performed during the cell search procedure so that the conventional cross-correlator can be used to evaluate the preamble time-of-arrival.

The remaining CFO error needs to be tracked during the data communication phase, together with the SCO and the phase noise.

Channel tracking

Compared to the WLAN communication systems operating in a quasi-static indoor environment, the cellular communication systems operate in a mobile outdoor environment so that the propagation channels need to be tracked before they can be equalized. To keep the low complexity property of the receiver, it is often assumed that the channel is constant during one OFDM block (even if it varies from one block to the next). Pilot symbols are usually inserted in the time/frequency resource grid defined by the successive OFDM blocks within one slot based on which the channel coefficients can be estimated and interpolated over the whole time and frequency domains.

In the case of a short cyclic prefix, the resource blocks defined in the 3GPP LTE system are composed of seven blocks of 12 sub-carriers. Four pilot symbols are inserted in the resource blocks so that the pilot overhead is less than 5% (see Figure 6.10).

When tracking the time-varying channel, the CFO remaining error and the SCO are also compensated. Therefore:

- no specific structure must be foreseen to track the CFO and SCO during the data communication;

- the CFO and SCO of the different user terminals are tracked independently (similarly to the user channels).

The common phase generated in currently developed analog front-end is usually negligible compared to the uncorrelated phase noise.

Unfortunately, the channel variations generate ICI within one OFDM symbol block that limit significantly the communication performance. The goal of the next section is to build advanced models of the ICI based on which the impact of the time variation can be better compensated.

6.2.4 Advanced channel tracking

A crucial property of the OFDM technique is the orthogonality of the modulation. Indeed, under ideal conditions, the sub-carriers are exactly orthogonal to each other or, equivalently, the sub-carriers lie exactly on the DFT grid. However, when the channel is time-varying, this orthogonality is lost. More precisely, the time-variations of the channel results in Doppler spread whereby the sub-carriers are no longer orthogonal to each other and a significant amount of ICI results, dramatically limiting the performance. It is interesting to note that some transceiver phenomena can be seen as generating ICI in the same way: the carrier and clock frequency offsets.

This degradation has two important consequences on the signal processing required at the receiver side:

- equalization cannot be implemented by a single tap equalization in the frequency domain because this does not take the interference from other sub-carriers into account;

- the channel estimation, usually performed by sending known symbols on (a sub-set of) the sub-carriers is not as straightforward because ICI again deteriorates the quality of the measured channel responses.

The equalization of multi-carrier systems in time-varying channels has been addressed by several authors. In Rugini *et al.* (2005b) and Rugini *et al.* (2005a), a band approximation of the frequency domain channel matrix is used to simplify the equalizer implementation. A banded linear MMSE equalizer and a banded decision feedback (DF) MMSE equalizer are designed based on the channel matrix approximation. Similarly, Schniter (2004) proposes to use windowing in the time domain to reduce the ICI. These solutions assume ideal channel knowledge. Schniter (2003) further proposes a low complexity channel estimation scheme, relying unfortunately on a specific arrangement of pilot symbols. Gorokhov and Linnartz (2004) use a Taylor approximation to model the channel time variations, but the associated maximum likelihood channel estimation has a high complexity. Finally, we propose an iterative channel estimator exploiting the knowledge of the ICI form when the Taylor approximation is limited to its first order (Bourdoux *et al.* 2006).

In this section, we show that the performance of the mobile 3GPP LTE system can be significantly improved by assuming that the channel varies linearly over a short burst of OFDM symbols. We build first a frequency domain matrix model, highlighting the banded structure of the channel matrix. We derive the ML channel estimator and linear MMSE equalizer (their simplification is out of the scope of this work). Afterwards, we demonstrate that the constraints on the CFO and SCO specifications are relaxed, making the design of the analog front-end and of the acquisition algorithms more realistic.

Frequency domain model

A simplified version of the OFDM system illustrated in Figure 2.1 is provided in Figure 6.13. Our model will be described for one OFDM symbol block but it should be clear that the OFDM frames consist of several OFDM symbol blocks. Since we assume in this section that there is no inter-block interference, this model for a single symbol block is sufficient. The model focuses on the convolution with the time-varying channel and on the conversion to the frequency domain.

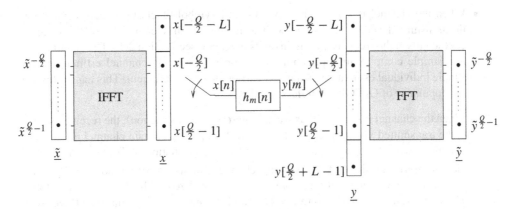

Figure 6.13 OFDM communications over a time-varying channel.

At each OFDM symbol time interval, a symbol vector $\tilde{\underline{x}}$ of size Q is transmitted over the channel. The frequency domain symbol vector $\tilde{\underline{x}}$ is converted by a size-Q inverse discrete

Fourier transform to a size-Q vector of time domain samples, which is then prepended with a cyclic prefix and parallel-to-serial converted, resulting in the discrete-time transmitted signal $x[n]$:

$$x[n] = \frac{1}{\sqrt{Q}} \sum_{q=-Q/2}^{Q/2-1} \tilde{x}^q e^{j2\pi(nq/Q)} \qquad (6.44)$$

for $n = -Q/2 - L, \ldots, Q/2 - 1$. The linear time-variant channel is modeled by the time-variant discrete impulse response $h_m[n]$ representing the response at time m to an impulse applied at time $m - n$ (in other words, m is the time dimension and n is the delay dimension). The received samples are given by:

$$y[m] = \sum_{n=-(Q/2)-L}^{(Q/2)-1} x[n] h_m[m-n] \qquad (6.45)$$

Note that the additive noise has been omitted for the sake of simplicity. After serial-to-parallel conversion, cyclic prefix removal and discrete Fourier transform, the elements of the frequency domain received signal vector $\tilde{y}$ are expressed as:

$$\tilde{y}^p = \frac{1}{\sqrt{Q}} \sum_{m=-(Q/2)}^{(Q/2)-1} y[m] e^{-j2\pi(mp/Q)} \qquad (6.46)$$

for $p = -Q/2, \ldots, Q/2 - 1$. Combining Equations (6.44), (6.45) and (6.46), we obtain:

$$\tilde{y}^p = \frac{1}{Q} \sum_{m=-(Q/2)}^{(Q/2)-1} \sum_{n=-(Q/2)-L}^{(Q/2)-1} \sum_{q=-(Q/2)}^{(Q/2)-1} \tilde{x}^q h_m[m-n] e^{j2\pi(nq-mp/Q)} \qquad (6.47)$$

The last expression (6.47) is usually compactly written in a matrix form. The result is different for the static channel case and for the time-varying channel case:

- When the channel is static during one OFDM symbol block, the received vector is the transmitted vector multiplied with a diagonal matrix composed of the channel coefficients in the frequency domain on its diagonal (see Section 2.1). The equalization is a simple complex multiplication per sub-carrier and the channel estimation is a simple individual estimation of the sub-carrier channel coefficient. This has motivated the popularity of OFDM in wireless standards.

- When the channel is not static (as we are interested in this section), the received vector is the transmitted vector multiplied with a full frequency domain channel matrix (the columns of the time domain channel matrix $\underline{\dot{H}}$ in Equation (2.7) are composed of the successive versions of the varying channel impulse response so that $\underline{\dot{H}}$ is not circulant anymore and can therefore not be diagonalized with FFT/IFFT operators). The equalization is much more involved and the number of channel coefficients to estimate becomes exceedingly large.

In order to facilitate the channel estimation and equalization, we approximate the channel time variation with a first order Taylor series (Gorokhov and Linnartz 2004):

$$h_m[n] = h_0[n] + m\dot{h}_0[n] \qquad (6.48)$$

where $h_0[n]$ is the channel impulse response at time 0 and $\dot{h}_0[n]$ is the channel impulse response first-order derivative at time 0. Including Equation (6.48) into (6.47), we obtain:

$$\tilde{y}^p = \frac{1}{Q} \sum_{m=-(Q/2)}^{(Q/2)-1} \sum_{n=-(Q/2)-L}^{(Q/2)-1} \sum_{q=-(Q/2)}^{(Q/2)-1} \tilde{x}^q (h_0[m-n] + m\dot{h}_0[m-n]) \, e^{j2\pi(nq-mp/Q)} \quad (6.49)$$

which can be simplified to:

$$\tilde{y}^p = \sqrt{Q}\tilde{h}_0^p \tilde{x}^p + \sum_{q=-(Q/2)}^{(Q/2)-1} \sqrt{Q}\tilde{\dot{h}}_0^q \left(\frac{1}{Q} \sum_{m=-(Q/2)}^{(Q/2)-1} m \, e^{j2\pi m(q-p)/Q} \right) \tilde{x}^q \quad (6.50)$$

in which $\tilde{h}_0^p$ is the channel frequency response at time 0 and $\tilde{\dot{h}}_0^p$ is the channel first-order derivative frequency response at time 0. It can finally be reorganized in a matrix model:

$$\underline{\tilde{y}} = \underline{\underline{\Lambda}}_{\tilde{h}} \cdot \underline{\tilde{x}} + \underline{\underline{\Phi}} \cdot \underline{\underline{\Lambda}}_{\tilde{\dot{h}}} \cdot \underline{\tilde{x}} \quad (6.51)$$

where $\underline{\tilde{x}}$ and $\underline{\tilde{y}}$ are the symbol and received vectors, respectively, $\underline{\underline{\Lambda}}_{\tilde{h}}$ and $\underline{\underline{\Lambda}}_{\tilde{\dot{h}}}$ are the diagonal matrices composed of the channel response and channel response first-order derivative in the frequency domain on their diagonal, respectively, and

$$\underline{\underline{\Phi}} := \left[\frac{1}{Q} \sum_{m=-(Q/2)}^{(Q/2)-1} m \, e^{j2\pi m(q-p)/Q} \right]_{p,q=-(Q/2),\ldots,(Q/2)} \quad (6.52)$$

It can be seen that the matrix $\underline{\underline{\Phi}}$ is a circulant matrix having as middle column the FFT of the vector $[-Q/2, \ldots, Q/2-1]^T / \sqrt{Q}$.

The first term of the matrix model (6.51) corresponds to the conventional (static) part of the channel (compare the result to the Equation (2.14)) whereas the second term provides the ICI term. The ICI term has a special structure that will be key to derive practical equalizer and channel estimation schemes: it is nearly banded in the sense that the elements having the highest amplitude are located on the main diagonal or in the sub-diagonals close to the main diagonal. The closed-form expression of the ICI term is especially useful since it allows the coefficients of the banded frequency doppler channel matrix to be computed, from which the equalizer and channel estimator can be computed.

Channel equalization

The simplest equalizers are linear equalizers where the transmitted symbol vector $\underline{\tilde{x}}$ is recovered by a multiplication of the received vector $\underline{\tilde{y}}$ expressed in Equation (6.51) with the equalizer matrix. Based on the system model (6.51) and exploiting the orthogonality between the error and the received signal, the MMSE equalizer can easily be derived (see Appendix A). It requires, however, a prohibitively complex inversion of the square channel auto-correlation matrix of size Q.

Different approximations of the channel matrix enable a significant reduction of the linear MMSE equalizer computation complexity.

- A band approximation of the channel matrix, consisting in taking only a subset of q sub-diagonals above and below the main diagonal and neglecting the other sub-diagonals, can first be performed. The channel auto-correlation matrix LDL factorization can then be simplified, reducing therefore the complexity of its inversion (Rugini *et al.* 2005b).

- The linear MMSE equalizer equalizes the complete OFDM symbol (i.e. all sub-carriers) at once. A closer look at the channel matrix reveals that, for a given sub-carrier, a local MMSE solution can be computed that involves only the neighboring sub-carriers according to the width $(2q + 1)$ of the band approximation. The complex inversion of the channel auto-correlation matrix of size Q reduces to Q inversions of much smaller matrices of size $2q$.

- The channel auto-correlation matrix can be seen as a dominant diagonal matrix plus a non-diagonal error matrix. Using a first-order Taylor approximation, the inverse is obtained by subtracting the weighted error from the diagonal matrix.

Other interesting equalizers are the nonlinear DF equalizers. By feeding back the already detected symbols to the equalizer, the interference that they generate on the received signal can be reconstructed and cancelled. Similarly to the linear MMSE equalizer, the complexity of their design can be significantly reduced by exploiting the band structure of the channel matrix (Rugini *et al.* 2005a).

Channel estimation

In order to estimate the transmitted vector of symbols $\tilde{\underline{x}}$ based on the received vector $\tilde{\underline{y}}$ given in Equation (6.51), both the channel transfer function and its linear variation over time should be known. They are usually estimated based on a grid of pilot symbols inserted in the burst of transmitted OFDM symbols (see, for example, the burst description for the 3GPP LTE system provided in Figure 6.11).

For the sake of simplicity, we assume a reference channel estimation burst composed as follows: one full OFDM pilot symbol is located at both sides of the burst, N OFDM data symbols are located in the center of the burst (see Figure 6.14). The size-L channel impulse response vector and its slope are denoted by $\underline{h}_i$ and $\underline{\dot{h}}_i$, respectively, where the index $i = 1, 2$ refers to the two pilot positions. Assuming that the channel varies linearly over the burst, it is clear that:

$$\underline{\dot{h}}_2 = \underline{\dot{h}}_1 \tag{6.53}$$

$$\underline{h}_2 = \underline{h}_1 + K \underline{\dot{h}}_1 \tag{6.54}$$

where $K = (Q + L_{cp})(N + 1)$ in with Q is the number of carriers, L_{cp} is the cyclic prefix length.

Since the length $L \leq L_{cp}$ of the channel impulse response and of its slope is much smaller than the block length Q, it is more efficient to estimate the channels in the time domain instead of the frequency domain ($L \ll Q$ parameters need to be estimated instead of Q). The channel frequency response (respectively, its slope) is simply the FFT of the channel impulse response (respectively, its slope) such that:

$$\underline{\tilde{h}}_i = \underline{\underline{F}} \cdot \underline{h}_i \tag{6.55}$$

$$\underline{\tilde{\dot{h}}}_i = \underline{\underline{F}} \cdot \underline{\dot{h}}_i \tag{6.56}$$

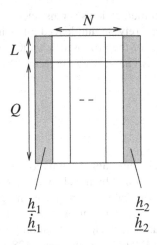

Figure 6.14 Reference channel estimation OFDM burst.

for $i = 1, 2$, where $\bar{\underline{F}}$ is composed of the L first columns of the FFT matrix of size Q. Therefore the matrix model (6.51) can be rewritten for each pilot instant as ($i = 1, 2$):

$$\underline{\tilde{y}}_i = \underline{\underline{\Lambda}}_{\tilde{x}_i} \cdot \bar{\underline{\underline{F}}} \cdot \underline{h}_i + \underline{\underline{\Phi}} \cdot \underline{\underline{\Lambda}}_{\tilde{x}_i} \cdot \bar{\underline{\underline{F}}} \cdot \dot{\underline{h}}_i + \underline{\tilde{z}}_i$$

in which $\underline{\tilde{z}}_i$ is the additional noise vector.

By stacking the received vectors corresponding to the two time instants on each other and relying on Equations (6.53) and (6.54), we obtain the composite matrix model:

$$\underline{y} = \underline{\underline{P}} \cdot \underline{h} + \underline{z} \qquad (6.57)$$

where the received and noise vectors are defined as:

$$\underline{y} := \begin{bmatrix} \underline{\tilde{y}}_1 & \underline{\tilde{y}}_2 \end{bmatrix}^T \qquad (6.58)$$

$$\underline{z} := \begin{bmatrix} \underline{\tilde{z}}_1 & \underline{\tilde{z}}_2 \end{bmatrix}^T \qquad (6.59)$$

the channel vector is defined as:

$$\underline{h} = \begin{bmatrix} \underline{h}_1 & \dot{\underline{h}}_1 \end{bmatrix}^T \qquad (6.60)$$

and the pilot matrix is defined as:

$$\underline{\underline{P}} := \begin{bmatrix} \underline{\underline{\Lambda}}_{\tilde{x}_1} \cdot \bar{\underline{\underline{F}}} & \underline{\underline{\Phi}} \cdot \underline{\underline{\Lambda}}_{\tilde{x}_1} \cdot \bar{\underline{\underline{F}}} \\ \underline{\underline{\Lambda}}_{\tilde{x}_2} \cdot \bar{\underline{\underline{F}}} & (K\underline{\underline{I}}_L + \underline{\underline{\Phi}}) \cdot \underline{\underline{\Lambda}}_{\tilde{x}_2} \cdot \bar{\underline{\underline{F}}} \end{bmatrix} \qquad (6.61)$$

The ML estimate of the channel vector $\underline{h}$ is simply given by (see Appendix B):

$$\hat{\underline{h}} = (\underline{\underline{P}}^H \cdot \underline{\underline{P}})^{-1} \cdot \underline{\underline{P}}^H \cdot \underline{y} \qquad (6.62)$$

Computing the ML estimator involves the inversion of a square matrix of size $2L$. Like the channel equalizer, it can be simplified by considering a banded approximation of the ICI matrix $\underline{\underline{\phi}}$.

On the other hand, actual communication systems, such as the 3GPP LTE, do not consider full pilot symbol blocks but rather rely on interleaved pilots in the data symbols of different OFDM blocks. The performance of the proposed scheme can be approached by iteratively reconstructing the ICI and by removing it from the received signal (Bourdoux *et al.* 2006).

Performance analysis

The goal of this section is to assess the channel estimation and equalization performance gain obtained when part of the ICI generated by the channel time variation is taken into account. We focus on the downlink of a cellular communication system operating in an outdoor suburban macro-cell propagation environment. A 2 GHz carrier frequency and a 1.25 MHz communication bandwidth are assumed. The modulation parameters are selected according to the 3GPP LTE standard: OFDM(A) air interface, 128 sub-carriers, 76 occupied sub-carriers, a cyclic prefix length equal to 32 (long cyclic prefix) and a 16QAM constellation. No channel coding is performed in this analysis. The time-varying channel realizations are generated according to the 3GPP TR25.996 geometrical channel model (3GPP channel model 2003).

Figure 6.14 illustrates the burst taken as a reference for this analysis. The channel is estimated based on two OFDM pilot blocks located on the borders of the burst and interpolated linearly over the OFDM data blocks located in the middle of the burst. Compared to the conventional channel estimation and equalization schemes that simply neglect the ICI generated by the channel time variation, the proposed channel estimation and equalization schemes assume rather that the ICI is generated by a linear channel variation over each OFDM block. Therefore, both systems suffer from the inaccurate knowledge of two elements: the static channel component of each OFDM block and the form of the ICI. In the following, we compare the performance of the conventional system that neglects the ICI (dashed curves) to the performance of the proposed system that supposes that the channel varies linearly (solid curves).

The influence of the mobile terminal speed on the system performance is first assessed. Figure 6.15 illustrates the channel estimation MSE and Figure 6.16 illustrates the overall system BER, respectively, for a varying terminal speed. The interpolation window N is fixed to 1 OFDM block. When the terminal is mobile, the performance floors to a minimum increasing with the terminal speed. Taking the part of the ICI corresponding to the channel linear variation into account brings a significant performance gain (a factor 10 in channel estimation MSE and a factor 4 in BER are observed at high SNR for a 130 km/h terminal speed).

The influence of the time interpolation window length N on the system performance is secondly assessed. Figure 6.17 illustrates the channel estimation MSE and Figure 6.18 illustrates the overall system BER, respectively, for a varying time interpolation window N. The terminal speed is fixed to 130 km/h. The performance of the conventional system is nearly independent of the interpolation window, demonstrating that the neglected ICI is the limiting factor. The performance of the proposed system depends heavily on the interpolation window (at least the final BER), showing that the linear channel variation approximation is less accurate when the time interval increases.

EMERGING WIRELESS COMMUNICATION SYSTEMS

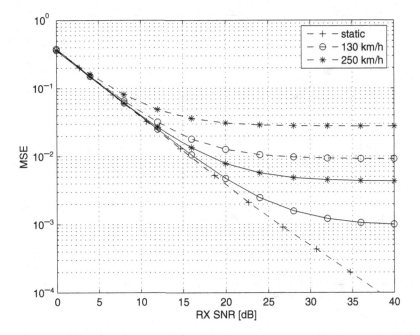

Figure 6.15 Impact of the speed on the channel estimation MSE. Dashed curves: ICI neglected; solid curves: linear approximation.

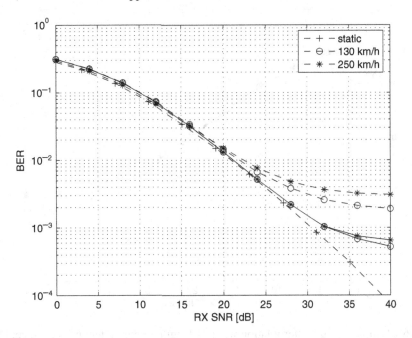

Figure 6.16 Impact of the speed on the system performance. Dashed curves: ICI neglected; solid curves: linear approximation.

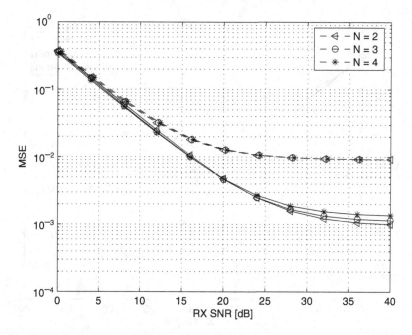

Figure 6.17 Impact of the interpolation window length on the channel estimation MSE. Dashed curves: ICI neglected; solid curves: linear approximation.

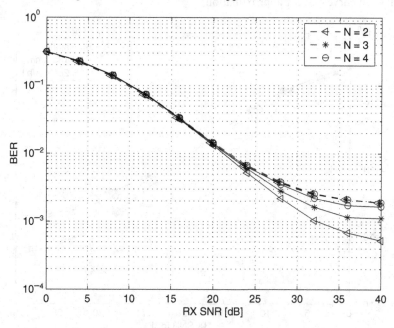

Figure 6.18 Impact of the interpolation window length on the system performance. Dashed curves: ICI neglected; solid curves: linear approximation.

EMERGING WIRELESS COMMUNICATION SYSTEMS

Improved sensitivity to the CFO

We first demonstrate analytically that small CFO values can be seen as a linearly time-varying channel. In the presence of CFO, the time-varying discrete impulse response $h_m[n]$ in Equation (6.45) takes the form:

$$h_m[n] = e^{j\phi m} h[n] \tag{6.63}$$

where $h[n]$ is the static discrete channel impulse response and ϕ is the phase drift due to the CFO over one sample. For small values of the CFO ($\phi \ll 1$), it is approximately equal to:

$$h_m[n] \simeq h[n] + j\phi m h[n] \tag{6.64}$$

which can be seen as a linearly time-varying channel as expressed in Equation (6.48) with:

$$h_0[n] = h[n] \tag{6.65}$$

$$h_1[n] = j\phi h[n] \tag{6.66}$$

Therefore, the channel linear variation is proportional to the CFO and to the channel response itself.

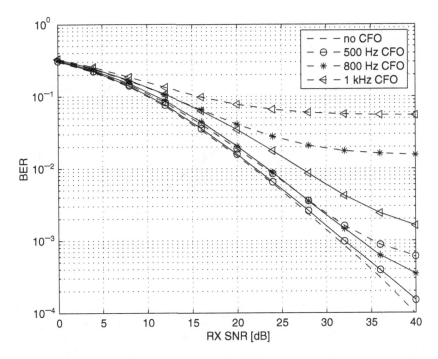

Figure 6.19 Impact of the CFO on the system performance; static terminal, interpolation window: 2 OFDM symbols. Dashed curves: ICI neglected, solid curves: linear approximation.

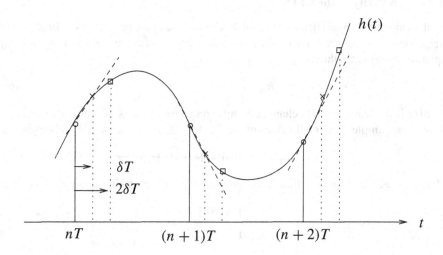

Figure 6.20 Series of discrete impulse responses $h_m[n]$ due to the SCO ($h_0[n]$: ○, $h_1[n]$: ×, $h_2[n]$: □, ...).

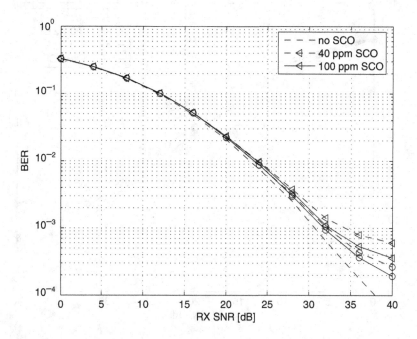

Figure 6.21 Impact of the SCO on the system performance; static terminal, 512 carriers, interpolation window: two OFDM symbols. Dashed curves: ICI neglected, solid curves: linear approximation.

Figure 6.19 illustrates the overall system BER for a varying receiver CFO. The terminal is static and the interpolation window N is fixed to 2 OFDM blocks. As we have demonstrated, the ICI caused by small values of the CFO can be well approximated by a linear channel variation so that the proposed system outperforms significantly the conventional system in the presence of CFO. The proposed system can afford up to 1 kHz CFO.

Improved sensitivity to the SCO

Secondly, we follow a similar approach to demonstrate that the small SCO values can also be seen as a linearly time-varying channel. In the presence of SCO, the time-varying discrete impulse response $h_m[n]$ in (6.45) takes the form:

$$h_m[n] = h(nT + m\delta T) \quad (6.67)$$

where $h(t)$ is the static continuous composite impulse response, including the transmit/receive filters and the propagation channels, and δ is the SCO. If $\dot{h}(t)$ denotes the first-order derivative of $h(t)$, $h_m[n]$ is approximately equal for small values of the SCO ($\delta \ll 1$) to:

$$h_m[n] \simeq h(nT) + m\delta T \dot{h}(nT) \quad (6.68)$$

which can be seen as a linearly time-varying channel as expressed in Equation (6.48) with:

$$h_0[n] = h(nT) \quad (6.69)$$

$$h_1[n] = \delta T \dot{h}(nT) \quad (6.70)$$

Therefore, the channel linear variation is proportional to the SCO and to the channel response first-order derivative. Figure 6.20 illustrates the series of discrete channel impulse responses $h_m[n]$ generated by the presence of SCO ($m = 0, 1, 2\ldots$) and shows that it can be approximated with a linear channel variation.

Figure 6.21 illustrates the overall system BER for a varying receiver SCO. The terminal is static and the interpolation window N is fixed to 2 OFDM blocks. As we have demonstrated, the ICI caused by the SCO can again be approximated by a linear channel variation so that the proposed system outperforms also the conventional system in the presence of SCO. However, the gain is less pronounced than in the case of the CFO (the impact of the SCO is also generally negligible).

References

3GPP channel model (2003) TR25.996 v6.1.0, spatial channel model for multiple input multiple output (MIMO) simulations.

Bourdoux, A., Horlin, F., Lopez-Estraviz, E. and Van der Perre, L. (2006) Practical channel estimation for OFDM in time varying channels. *IEEE Proceedings of Globecom*, pp. 1–5.

Gorokhov, A. and Linnartz, J. (2004) Robust OFDM receiver for dispersive time-varying channels: Equalization and channel acquisition. *IEEE Transactions on Communications* 52(4), 572–583.

Horlin, F. and Van der Perre, L. (2004) Optimal training sequences for low complexity ML multi-channel estimation in multi-user MIMO OFDM-based communications. *IEEE Proceedings of International Conference on Communications*, vol. 4, pp. 2427–2431.

IEEE802-11b (1999) IEEE802.11b-1999, wireless lan medium access control (MAC) and physical layer (PHY) specifications, high-speed physical layer in the 2.4 GHz band http://www.ieee802.org/11/.

IEEE802.11 (1999) Wireless LAN medium access control (MAC) and physical layer (PHY) specifications http://www.ieee802.org/11.

IEEE802.11a (1999) IEEE802.11a-1999, wireless lan medium access control (MAC) and physical layer (PHY) specifications, high-speed physical layer in the 5 GHz band http://www.ieee802.org/11/.

IEEE802.11g (2003) IEEE802.11g-2003, wireless lan medium access control (MAC) and physical layer (PHY) specifications, amendment 4: Further higher data rate extension in the 2.4 GHz band http://www.ieee802.org/11/.

IEEE802.11n (2007) IEEE802.11n/D3.01, wireless lan medium access control (MAC) and physical layer (PHY) specifications, amendment 4: Enhancements for higher throughput http://www.ieee802.org/11/.

Kaiser, T., Bourdoux, A., Boche, H., Fonollosa, J., Andersen, J. and Utschick, W. (2005) *Smart Antennas – State of the Art*. Eurasip.

Liu, J., Vandersteen, G., Craninckx, J., Libois, M., Wouters, M., Petré, F. and Barel, A. (2006) A novel and low-cost analog front-end mismatch calibration scheme for MIMO-OFDM WLANs. *IEEE Proceedings of Radio and Wireless Symposium*, pp. 219–222.

Meyr, H., Moeneclaey, M. and Fechtel, S. (1998) *Digital Communication Receivers Synchronization, Channel Estimation and Signal Processing*. John Wiley & Sons.

Rugini, L., Banelli, P. and Leus, G. (2005a) Block DFE and windowing for doppler-affected OFDM systems. *IEEE Proceedings of Signal Processing Advances in Wireless Communications*, pp. 470–474.

Rugini, L., Banelli, P. and Leus, G. (2005b) Reduced complexity equalization of MC-CDMA systems over time-varying channels. *IEEE Proceedings of International Conference on Acoustics, Speech and Signal Processing*, vol. 3, pp. 473–476.

Schniter, P. (2003) Low-complexity channel estimation of doubly selective channels. *IEEE Proceedings of Signal Processing Advances in Wireless Communications*, pp. 200–204.

Schniter, P. (2004) Low complexity equalization of OFDM in doubly selective channels. *IEEE Transactions on Signal Processing* **52**(4), 1002–1011.

Appendices

A

MMSE Linear Detector

A usual problem in the design of a digital communication system is to estimate a vector of symbols $\underline{d}$ of size N_d transmitted at one side of the link based on the observation of a vector of samples $\underline{r}$ of size N_r received at the other side of the link. The received vector is typically a channel distorted version of the transmitted vector plus an additive Gaussian noise, so that:

$$\underline{r} = \underline{\underline{G}} \cdot \underline{d} + \underline{n} \tag{A1}$$

in which $\underline{\underline{G}}$ of size $N_r \times N_d$ models the channel distortion and $\underline{n}$ of size N_r is the additive noise. The rank of the matrix $\underline{\underline{G}}$ should be at least equal to N_d. When a linear joint detector $\underline{\underline{F}}$ of size $N_d \times N_r$ is applied on the received signal (A1) in order to estimate the transmitted symbols, the expression of the estimated vector is:

$$\underline{\hat{d}} = \underline{\underline{F}} \cdot \underline{r} \tag{A2}$$

and the prediction error vector is defined as:

$$\underline{\epsilon} := \underline{d} - \underline{\hat{d}} \tag{A3}$$

The linear MMSE detector is designed to minimize the variance of the symbol estimation errors, or equivalently the diagonal elements of the symbol estimation error auto-correlation matrix defined as $\underline{\underline{R}}_{\epsilon\epsilon} := \mathcal{E}[\underline{\epsilon} \cdot \underline{\epsilon}^H]$. In order to minimize the trace of $\underline{\underline{R}}_{\epsilon\epsilon}$, the orthogonality principle is used (Trees 1971). It states that the optimal estimator of a random variable is obtained when the error is orthogonal to the observation. Hence the cross-correlation between the error and received vectors is:

$$\underline{\underline{R}}_{\epsilon r} = \underline{\underline{0}}_{N_d \times N_r} \tag{A4}$$

The expression of the detector becomes (Klein *et al.* 1996):

$$\underline{\underline{F}} = \underline{\underline{R}}_{dr} \cdot \underline{\underline{R}}_{rr}^{-1} \tag{A5}$$

$$= \underline{\underline{R}}_{dd} \cdot \underline{\underline{G}}^H [\underline{\underline{G}} \cdot \underline{\underline{R}}_{dd} \cdot \underline{\underline{G}}^H + \underline{\underline{R}}_{nn}]^{-1} \tag{A6}$$

$$= [\underline{\underline{R}}_{dd}^{-1} + \underline{\underline{G}}^H \cdot \underline{\underline{R}}_{nn}^{-1} \cdot \underline{\underline{G}}]^{-1} \cdot \underline{\underline{G}}^H \cdot \underline{\underline{R}}_{nn}^{-1} \tag{A7}$$

where Equation (A5) comes from the orthogonality principle (A4) in which the prediction error (A3) and the symbol estimate (A2) are introduced. The initial model (A1) is used to obtain Equation (A6). Finally, the expression of the MMSE joint detector (A7) comes from the matrix inversion lemma.[1] The linear MMSE joint detector is composed of the whitening matched filter $\underline{\underline{G}}^H \cdot \underline{\underline{R}}_{nn}^{-1}$ that maximizes the SNR, followed by the multiplication with a square matrix that minimizes the remaining interference (Klein et al. 1996; Vandendorpe 1997). The symbol estimate (A2) is:

$$\hat{\underline{d}} = \underline{\underline{F}} \cdot \underline{\underline{G}} \cdot \underline{d} + \underline{\underline{F}} \cdot \underline{n} \tag{A8}$$

$$= [\underline{\underline{R}}_{dd}^{-1} + \underline{\underline{G}}^H \cdot \underline{\underline{R}}_{nn}^{-1} \cdot \underline{\underline{G}}]^{-1} \underline{\underline{G}}^H \cdot \underline{\underline{R}}_{nn}^{-1} \cdot \underline{\underline{G}} \underline{d}$$

$$+ [\underline{\underline{R}}_{dd}^{-1} + \underline{\underline{G}}^H \cdot \underline{\underline{R}}_{nn}^{-1} \cdot \underline{\underline{G}}]^{-1} \underline{\underline{G}}^H \cdot \underline{\underline{R}}_{nn}^{-1} \cdot \underline{n} \tag{A9}$$

$$= \left(\underline{\underline{I}}_{N_d} - [\underline{\underline{R}}_{dd}^{-1} + \underline{\underline{G}}^H \cdot \underline{\underline{R}}_{nn}^{-1} \cdot \underline{\underline{G}}]^{-1} \underline{\underline{R}}_{dd}^{-1}\right) \underline{d}$$

$$+ [\underline{\underline{R}}_{dd}^{-1} + \underline{\underline{G}}^H \cdot \underline{\underline{R}}_{nn}^{-1} \cdot \underline{\underline{G}}]^{-1} \cdot \underline{\underline{G}}^H \cdot \underline{\underline{R}}_{nn}^{-1} \cdot \underline{n} \tag{A10}$$

and the corresponding error auto-correlation matrix becomes:

$$\underline{\underline{R}}_{\epsilon\epsilon} = [\underline{\underline{R}}_{dd}^{-1} + \underline{\underline{G}}^H \cdot \underline{\underline{R}}_{nn}^{-1} \cdot \underline{\underline{G}}]^{-1} \tag{A11}$$

The error variance of any symbol estimate can be found at the appropriate location on the diagonal of the matrix $\underline{\underline{R}}_{\epsilon\epsilon}$.

If the noise power becomes negligible as compared to the signal power, the MMSE joint detector reduces to the ZF joint detector that simply inverts the channel matrix $\underline{\underline{G}}$.

References

Klein, A., Kaleh, G.K. and Baier, P.W. (1996) Zero forcing and minimum mean square error equalization for multiuser detection in code division multiple access channels. *IEEE Transactions on Vehicular Technology* **45**(2), 276–287.

Van Trees, H.L. (1971) *Detection, Estimation, and Modulation Theory*. J. Wiley & Sons.

Vandendorpe, L. (1997) Performance analysis of IIR and FIR linear and decision-feedback MIMO equalizers for transmultiplexers. *IEEE Proceedings of International Conference on Communications*, vol. 2, pp. 657–661, Montreal, Canada.

---

[1] If $\underline{\underline{X}} = \underline{\underline{P}}^{-1} + \underline{\underline{Q}} \underline{\underline{R}}^{-1} \underline{\underline{Q}}^H$ in which the matrices $\underline{\underline{X}}$, $\underline{\underline{P}}$ and $\underline{\underline{R}}$ are square positive-definite, then $\underline{\underline{X}}^{-1} = \underline{\underline{P}} - \underline{\underline{P}} \cdot \underline{\underline{Q}} \cdot (\underline{\underline{R}} + \underline{\underline{Q}}^H \cdot \underline{\underline{P}} \cdot \underline{\underline{Q}})^{-1} \underline{\underline{Q}}^H \cdot \underline{\underline{P}}$.

B

ML Channel Estimator

The channel equalization methods integrated in most of the digital communication systems usually rely on the knowledge of the channel impulse response. Pilot symbols are often interleaved in the data symbols based on which the channel impulse response can be estimated. A matrix model is typically built that links the channel impulse response vector $\underline{h}$ of size L to the observed vector of samples $\underline{r}$ of size N_r:

$$\underline{r} = \underline{\underline{P}} \cdot \underline{h} + \underline{n} \tag{B1}$$

in which $\underline{\underline{P}}$ of size $N_r \times L$ is a matrix composed of the transmitted pilot symbols and $\underline{n}$ of size N_r is the additive noise composed of independent Gaussian elements of variance σ_n^2.

The ML estimate $\underline{\hat{h}}$ of the channel impulse response vector $\underline{h}$ is selected such that:

$$\underline{\hat{h}} = \max_{\underline{h}}[p(\underline{r}|\underline{h})] \tag{B2}$$

$$= \max_{\underline{h}}[-2\Re(\underline{r}^H \cdot \underline{\underline{P}} \cdot \underline{h}) + \underline{h}^H \cdot \underline{\underline{P}}^H \cdot \underline{\underline{P}} \cdot \underline{h}] \tag{B3}$$

where the second equality has been obtained by taking the expression of the noise probability into account. After optimization, the ML estimate is given by Deneire et al. (2003) and Kay (1993):

$$\underline{\hat{h}} = (\underline{\underline{P}}^H \cdot \underline{\underline{P}})^{-1} \cdot \underline{\underline{P}}^H \cdot \underline{r} \tag{B4}$$

$$= \underline{h} + (\underline{\underline{P}}^H \cdot \underline{\underline{P}})^{-1} \cdot \underline{\underline{P}}^H \cdot \underline{n} \tag{B5}$$

where it can be seen that the ML estimate is unbiased. The error auto-correlation matrix is finally given by:

$$\underline{\underline{R}}_{h_\epsilon h_\epsilon} = \mathcal{E}[(\underline{h} - \underline{\hat{h}}) \cdot (\underline{h} - \underline{\hat{h}})^H] \tag{B6}$$

$$= \sigma_n^2 (\underline{\underline{P}}^H \cdot \underline{\underline{P}})^{-1} \tag{B7}$$

References

Deneire, L., Vandenameele, P., Van der Perre, L., Gyselinck, B. and Engels, M. (2003) A low complexity ML channel estimator for OFDM. *IEEE Transactions on Communications* **51**(2), 135–140.

Kay, S.M. (1993) *Fundamentals of Statistical Signal Processing: Estimation Theory.* Prentice-Hall International Editions.

C

Matlab Models of Non-Idealities

C.1 Receiver non-idealities

C.1.1 Global RX non-idealities

```
% ===========================================================================
% Book title: Digital Front-End Compensation for Emerging Wireless Systems
% Authors:    Francois Horlin (ULB), Andre Bourdoux (IMEC)
% Editor:     John Wiley & Sons, Ltd
% Date:       2008
% ===========================================================================
%
% DESCRIPTION
%   This script provides an example to use the front-end RX non-ideality
%   functions.
%
% INPUTS        none
%
% OUTPUTS
%   - in        ideal input signal
%   - out       output signals affected by non-idealities
%   - a         structure containing all non-ideality parameters
%                   AWGN: [1x1 struct]
%                    CFO: [1x1 struct]
%                     PN: [1x1 struct]
%                     IQ: [1x1 struct]
%                    AGC: [1x1 struct]
%                    SCO: [1x1 struct]
%                    CNQ: [1x1 struct]

clear
% This generates a hypothetical received signal
Ns = 10000;
in = 5*exp(j*2*pi*(1:Ns)/128).';      % in must be a column vector

% We create variable "a", which is a structure containing
% all non-ideality parameters
```

```
% Use these flags to enable/disable the corresponding functions
a.AWGN.on = 1;
a.CFO.on  = 1;
a.PN.on   = 1;
a.IQ.on   = 1;
a.AGC.on  = 1;
a.SCO.on  = 1;
a.CNQ.on  = 1;

% Parameter settings
a.AWGN.snr = 15;        % SNR in dB
a.AWGN.osf = 4;         % over-sampling factor
a.AWGN.r_c = 0;         % 1--real, 2--complex

a.CFO.fs  = 80e6;       % sampling frequency
a.CFO.cfo = 40e-6;      % value of CFO

a.PN.dBc    = -20;      % integrated phase noise in dBc
a.PN.cutoff = 100e3;    % cutoff frequency of PSD
a.PN.floor  = -120;     % phase noise floor in dBc/Hz
a.PN.fs     = 80e6;     % sampling frequency

a.IQ.eps    = 0.03;     % amplitude imbalance
a.IQ.dphi   = 0.03;     % phase imbalance
a.IQ.random = 0;        % 0==fixed, 1==random

a.AGC.rms = 2;          % target RMS amplitude value
a.AGC.n1  = 1;          % first sample index for amplitude estimation
a.AGC.n2  = NaN;        % last sample index for amplitude estimation

a.SCO.sco = 40e-6;      % value of SCO

a.CNQ.enob = 6;         % effective number of bits
a.CNQ.CL   = 2.5;       % clip level

% Apply all non-idealities in order
out = in;   % This is needed in case no non-idealities are applied

% *** Additive white Gaussian noise ***
if (a.AWGN.on == 1)
    out = RX_AWGN(out,a.AWGN.snr,a.AWGN.osf,a.AWGN.r_c);
end

% *** Carrier frequency offset (CFO) ***
if (a.CFO.on == 1)
    out = RX_CFO(out,a.CFO.fs,a.CFO.cfo);
end

% *** ADD PHASE NOISE (PN) ***
if (a.PN.on == 1)
    out = RX_PN(out,a.PN.dBc,a.PN.cutoff,a.PN.floor,a.PN.fs);
end

% *** I/Q mismatches ***
if (a.IQ.on == 1)
   out = RX_IQ(out,a.IQ.eps,a.IQ.dphi,a.IQ.random);
end
```

MATLAB MODELS OF NON-IDEALITIES

```
% *** AGC ***
if (a.AGC.on == 1)
    out = RX_AGC(out,a.AGC.rms,1,1000);
end

% *** Sampling clock offset (SCO) ***
if (a.SCO.on == 1)
    out = RX_SCO(out,a.SCO.sco);
end

% *** CLIP AND QUANTIZE ***
if(a.CNQ.on == 1)
    out = RX_CNQ(out,a.CNQ.enob,a.CNQ.CL);
end
```

C.1.2 Receiver noise

```
function [ out ] = RX_AWGN(in, SNR_dB, OSF, R_C)
% ==========================================================================
% Book title: Digital Front-End Compensation for Emerging Wireless Systems
% Authors:    Francois Horlin (ULB), Andre Bourdoux (IMEC)
% Editor:     John Wiley & Sons, Ltd
% Date:       2008
% ==========================================================================
%
% DESCRIPTION
%   This function adds white Gaussian noise to the signal stream(s). The
%   same noise variance is added to each stream (i.e. antenna). The noise
%   variance is determined by the SNR, expressed in dB. The noise variance
%   is multiplied by the oversampling factor, so that the SNR in the signal
%   bandwidth is not modified by the oversampling. It is possible to have
%   real or complex input signal. This function assumes that the signal
%   variance is equal to 1. The noise variance is calculated accordingly.
%
% INPUTS
%   - in         [length x Ns]: Ns streams of input signals
%   - SNR_dB     [1]: SNR in dB
%   - OSF        [1]: over-sampling factor
%   - R_C        [1]: real or complex  (1==real, 0==complex)
%
% OUTPUTS
%   - out        [length x Ns]: Ns streams of output signals

if nargin==3
    R_C = 0;, disp('!!! complex signal is assumed !!!')
end
if nargin==2
    R_C = 0;, OSF =1;,  disp('!!! complex signal and OSF=1 is assumed !!!')
end

SNR = 10^(SNR_dB/10);     % Converts to linear SNR

if R_C
    % real signal, variance of signal assumed equal to 1
    noise = sqrt(OSF/(SNR)) * ( randn(size(in)) + j*randn(size(in)) );
else
    % complex signal, variance of signal assumed equal to 1
    noise = sqrt(OSF/(2*SNR)) * ( randn(size(in)) + j*randn(size(in)) );
```

```
end

out = in + noise;
```

C.1.3 Carrier frequency offset

```
function [out] = RX_CFO(in, FS, CFO);
% ==========================================================================
% Book title: Digital Front-End Compensation for Emerging Wireless Systems
% Authors:    Francois Horlin (ULB), Andre Bourdoux (IMEC)
% Editor:     John Wiley & Sons, Ltd
% Date:       2008
% ==========================================================================
%
% DESCRIPTION
%   This function applies CFO to the signal stream(s). Each column contains
%   one stream. If only one CFO value is given, it is applied to all
%   streams. Otherwise a vector of CFO values must be provided as input,
%   having a length equal to the number of streams. The sample rate must
%   also be provided.
%
% INPUTS
%    - in          [length x Ns]: input signal where each column is a signal
%                                 from a different antenna.
%    - FS          [1]: Sample rate
%    - CFO  [1] or [1 x length]: carrier frequency offset
%
% OUTPUTS
%    - out         [length x Ns]: Ns streams of output signals

[M N] = size(in);
if(length(CFO) ~= N & length(CFO) ~= 1)
    error('# of elements in CFO must be equal to 1 or to # of sequences');
end
if(length(CFO) == 1)
    CFO = repmat(CFO,1,size(in,2));
end

t = [0:1:M-1].'*1/FS;

vect = zeros(M,1);
out  = zeros(M,N);
for u = (1:N)
    vect = exp(j*2*pi*CFO(u)*t);
    out(:,u) = in(:,u).*vect;
end
```

C.1.4 Phase noise

```
function out = RX_PN(in,dBc, cutoff, floor, Fs)
% ==========================================================================
% Book title: Digital Front-End Compensation for Emerging Wireless Systems
% Authors:    François Horlin (ULB), André Bourdoux (IMEC)
% Editor:     John Wiley & Sons, Ltd
% Date:       2008
% ==========================================================================
%
```

MATLAB MODELS OF NON-IDEALITIES

```
% DESCRIPTION
%   This function applies phase noise to the signal stream(s). Each row
%   contains one stream. The same phase noise is applied to each stream as
%   if they are working with the same local oscillator. The phase noise has
%   a PSD defined by four values: integrated phase noise, cutoff frequency, a
%   phase noise floor and the sample rate. At the cutoff frequency, the PSD
%   starts a -20dB/dec roll-off
%
% INPUTS
%   - in        [length x Ns]: input signal where each row is a signal
%                               from a different antenna.
%   - dBc           [1]: integrated phase noise
%   - cutoff        [1]: cutoff frequency of the PSD
%   - floor         [1]: phase noise floor
%   - Fs            [1]: sample rate
%
% OUTPUTS
%   - out       [length x Ns]: Ns streams of output signals

[M N] = size(in);
ltx = 2^(fix(log2(M-0.5))+2);    % ltx will always be at least 2xM
if cutoff/(Fs/ltx)<16,
    ltx = 16*Fs/cutoff;
    ltx = 2^(fix(log2(ltx-0.5))+1);
end
PhaseNoise = FreqSynthLorenzian_new(dBc, cutoff, floor, Fs, ltx).';

Nphn = length(PhaseNoise);
if Nphn<M
    error(['Phase noise vector must be longer than ' num2str(Nsamples)])
end
Nstart = fix(rand(1,1)*(Nphn-M));   % random staring point
PhaseNoise = PhaseNoise(Nstart:Nstart+M-1);

% Add phase noise to data
out = zeros(size(in));
for k=1:N
    out(:,k) = in(:,k).*PhaseNoise.';
end
% -----------------------------------------------------

function y = FreqSynthLorenzian_new( K, B, p2, Fs, ltx);
%   - lo: time domain vector, possible time domain representation of the
%           oscillator amplitude.
%   - K: total noise power in dBc
%   - B: 3dB bandwidth
%   - p2: noise plateau level in dBc/Hz, to be set at #20dB above the noise
%           floor.
%   - ltx: length of the signal lo vector to return

% The negative frequencies of the BB equivalent LO spectrum are the
% complex conjugate of the positive frequencies (as a result, phi(t) is
% real). The reason for this is not the need for the phi(t) or lo(t) to be
% real, but it is coming from the analogy to FM or PM modulations that do
% create such 'symetric' frequency responses.
% - PHI(f), hence LO(f), should not be generated as a spectrum with a wanted
% shape, but as a PSD (representation for a random process) with a wanted
% mask.
```

```
global d
Ns  = ltx;               % Number of samples for ifft calculation
df  = Fs/Ns;             % frequency resolution
k   = [0:1:Ns/2-1, -Ns/2:1:-1];  % frequency index range
% frequency range: f = k * df
p2o = p2;
p2  = 10^(p2/10);        % in V^2/Hz
p2  = p2*df;             % in V^2/df

Ko  = K;
K   = 10^(K/10);         % in V^2
B   = B/df;              % 3-dB bandwidth index
% Low frequency part is defined by Lorenzian function
SSBmask = sqrt( K*B/pi./([1:Ns/2].^2+B^2) );
% High frequency part is defined by the noise floor of the system p2
SSBmask = max(SSBmask, sqrt(p2)*ones(size(SSBmask)));

%-- Phase noise PHI(f, f>0) is first generated as a wide band signal
PHI = sqrt(0.5) * abs( randn(1,Ns/2) + j*randn(1,Ns/2) );
PHI = PHI .* exp(j*2*pi*rand(1,Ns/2));
%-- Phase noise PHI(f, f>0) is shaped according to the wanted mask
PHI = PHI .* SSBmask;
%-- Phase noise PHI(f) is then generated from PHI(f, f>0)
% (no phase noise on the carrier)
PHI = [0 PHI(1:Ns/2-1) conj(PHI(Ns/2:-1:1))];
NoisePower = 10*log10( sum(abs(PHI).^2) );

if (0)
    f       = k(2:Ns/2)*df;
    PHI_f   = 20*log10(abs(PHI(2:Ns/2))/sqrt(df));
    SSBmask_f = 20*log10(SSBmask(1:Ns/2-1)/sqrt(df));
    % normalisation to get PHI(f) in dBc/Hz, and not dBc/df
    figure(20),semilogx(f, PHI_f, 'b-','linewidth',1);
    hold on; grid on; zoom on;
    semilogx(f, SSBmask_f, 'r-','linewidth',2);
    axis([f(1) f(end) p2o-10 SSBmask_f(1)+10]);
end;

%-- Correction for the integrated phase noise power:
PHI = PHI * 10^((Ko-NoisePower)/20);

%-- Phase noise phi(t) in the time domain
phi = ifft(PHI,Ns)*Ns;

%-- Local oscillator signal lo(t) in the time domain
lo = exp(j*phi);

%-- Local oscillator signal LO(f) in the frequency domain
if (0)
    LO = fft(lo,Ns)/Ns;

    f       = k(2:Ns/2)*df;
    LO_f    = 20*log10(abs(LO(2:Ns/2))/sqrt(df));
    SSBmask_f = 20*log10(SSBmask(1:Ns/2-1)/sqrt(df));
    figure(21);semilogx(f, LO_f, 'k-','linewidth',1);
    hold on; grid on; zoom on;
    semilogx(f, SSBmask_f, 'r-','linewidth',2);
    axis([f(1) f(end) p2o-10 SSBmask_f(1)+10]);
```

MATLAB MODELS OF NON-IDEALITIES   237

```
end;

%-- Prepare lo(t) for efficient use in dbbm_fe
y = lo(1,1:ltx*fix(Ns/ltx));
y = reshape(y,ltx,fix(Ns/ltx));
```

C.1.5 AGC

```
function [out, gain] = RX_AGC(in,RMS,n1,n2)
% ==========================================================================
% Book title: Digital Front-End Compensation for Emerging Wireless Systems
% Authors:       Francois Horlin (ULB), Andre Bourdoux (IMEC)
% Editor:        John Wiley & Sons, Ltd
% Date:          2008
% ==========================================================================
%
% DESCRIPTION
%   This function performs an automatic gain control on the signal
%   stream(s). Each column contains one stream. The target RMS value must be
%   passed. If only one RMS value is given, it is applied to all streams.
%   Otherwise a vector of RMS values must be provided as input, having a
%   length equal to the number of streams. Optionally, the index of the
%   start and end of the part on which to perform the amplitude estimation
%   can be provided. The output signal will be scaled in amplitude so that
%   its mean power is RMS^2. The gain that has been applied on each stream
%   is also returned.
%
% INPUTS
%   - in         [length x Ns]: input signal where each column is a signal
%                                from a different antenna.
%   - RMS    [1] or [1 x length]: target RMS value per streamSample rate
%   - n1                    [1]: optional starting point in the stream
%   - n2                    [1]: optional end point in the stream
%
% OUTPUTS
%   - out        [length x Ns]: Ns streams of output signals

[M N] = size(in);
if nargin==2, n1=1;, n2=N;, end
if(length(RMS) ~= N & length(RMS) ~= 1)
    error('# of elements in CFO must be equal to 1 or to # of sequences');
end
if(length(RMS) == 1)
    RMS = repmat(RMS,1,size(in,2));
end

out = zeros(M,N);
for k=1:N
    pwr(k) = in(n1:n2,k)'*in(n1:n2,k)/(n2-n1+1);
    out(:,k) = in(:,k)*RMS(k)/sqrt(pwr(k));
end
gain = RMS./sqrt(pwr);
```

C.1.6 Receive IQ imbalance

```
function out = RX_IQ(in,eps,dphi,random);
% ==========================================================================
```

```
% Book title: Digital Front-End Compensation for Emerging Wireless Systems
% Authors:    Francois Horlin (ULB), Andre Bourdoux (IMEC)
% Editor:     John Wiley & Sons, Ltd
% Date:       2008
% =========================================================================
%
% DESCRIPTION
%   This function applies IQ imbalance to the signal stream(s). Each column
%   contains one stream. If only one EPS or DPHI value is given, it is
%   applied to all streams. Otherwise a vector of EPS values and a vector of
%   DPHI values must be provided as input, each having a length equal to the
%   number of streams. The RANDOM parameter determines if EPS and DPHI are
%   used as such used or if random numbers are generated. In this latter case,
%   the amplitude is Gaussian N(0,EPS^2) and DPHI is uniformly distributed
%   in the interval [-DPHI,DPHI].
%
% INPUTS
%    - in          [length x Ns]: input signal where each column is a signal
%                                 from a different antenna.
%    - EPS    [1] or [1 x length]: amplitude imbalance
%    - DPHI   [1] or [1 x length]: phase imbalance
%    - random              [1]: 1 --> random value, 0 --> fixed value
%
% OUTPUTS
%    - out         [length x Ns]: Ns streams of output signals
[M N] = size(in);
[P,Q] = size(eps);
[R,S] = size(dphi);

if P~=1 & Q~=1, error('eps must be scalar or vector'),end
if P==1 & Q==1, eps=repmat(eps,1,N);,
elseif max(P,Q)==N, eps  = reshape(eps,1,N);
else  error('eps length must be equal to # streams')
end
if R~=1 & S~=1, error('dphi must be scalar or vector'),end
if R==1 & S==1, dphi=repmat(dphi,1,N);,
elseif max(R,S)==N, dphi = reshape(dphi,1,N);
else  error('dphi length must be equal to # streams')
end

if random
    eps  = eps.*randn(1,N);        % linear value (not in %)
    dphi = dphi*2.*(rand(1,N)-0.5); % in radians
else
    eps  = eps.*ones(1,N);                  % linear value (not in %)
    dphi = dphi.*ones(1,N);                 % in radians
end
for aa=1:N
    alph = cos(dphi(aa))-j*sin(dphi(aa))*eps(aa);
    beta = eps(aa)*cos(dphi(aa))+j*sin(dphi(aa));
    out(:,aa) = alph*in(:,aa)+beta*conj(in(:,aa));
end
```

C.1.7 Sampling clock offset

```
function out = RX_SCO(in,SCO);
% =========================================================================
% Book title: Digital Front-End Compensation for Emerging Wireless Systems
```

MATLAB MODELS OF NON-IDEALITIES

```
% Authors:    Francois Horlin (ULB), Andre Bourdoux (IMEC)
% Editor:     John Wiley & Sons, Ltd
% Date:       2008
% =========================================================================
%
% DESCRIPTION
% This function applies SCO to the signal stream(s). Each column contains
% one stream. The same SCO is applied to all streams. The SCO is provided
% as a relative offset with respect to the current sample rate.
%
% INPUTS
%   - in        [length x Ns]: input signal where each column is a signal
%                              from a different antenna.
%   - SCO       [1]: relative sampling clock offset
%                    wrt current sample rate.
%
% OUTPUTS
%   - out       [length x Ns]: Ns streams of output signals

[M N] = size(in);
if(SCO ~= 0)
    indx = (1:M)*(1+SCO);
    if SCO > 0
        in = [in; zeros(fix(indx(end))-M,N)];    % zero-padding: extra
                                                 % samples are needed
        for kk=1:N
            in2(:,kk) = lagrange_interp(in(:,kk),indx,'cubic');
        end
    else
        in = [zeros(1,N); in];
        indx = indx+1;
        for kk=1:N
            in2(:,kk) = lagrange_interp(in(:,kk),indx,'cubic');
        end
    end
    out = in2;
else
    out = in;
end
% ----------------------------------------------------

function yi = lagrange_interp(y,xi,interptype)

% Check arguments
if(nargin ~= 3)
    error('fast_interp expects 3 arguments');
end

if( (xi > length(y)) | (xi < 1) )
    error('xi must be between 1 and the length of y');
end

if(~ischar(interptype))
    error('Interpolation type must be a string');
end

% For linear interpolation, just use matlab's linear
% interpolation function
```

```
if(strcmp(interptype,'linear') == 1)
   yi = interp1(1:length(y),y,xi,'*linear');

% For cubic interpolation, calculate piecewise lagrange
% polynomial at specified points
elseif(strcmp(interptype,'cubic') == 1)
   yi = cubic_lagrange(y,xi);

% Otherwise print an error to the screen
else
   error('interptype must be either linear or cubic');

end
% --------------------------------------------------

function yi = cubic_lagrange(y,xi);
% CUBIC_LAGRANGE
%
% Calculates the values f(xi) at fractional indices xi using
% piecewise Lagrange polynomials. f(xi) is an approximation of y
% where each sample in y is assumed to be evenly spaced.

% add zeros to y to handle endpoints
y = reshape(y,length(y),1); % make sure y is a column vector
y(end+1:end+2) = 0;
y = [0;y];
xi = xi + 1;

% Get fractional part of indices
mu = xi - floor(xi);

% Get integer part of indices
xi = floor(xi);

% Get values from y used for cubic interpolation
xi = reshape([y(xi+2) y(xi+1) y(xi) y(xi-1)],length(mu),4);

% Perform interpolation using piecewise Lagrange polynomials
yi = sum((xi.*lagrange_coeff(3,mu.')).');

% --------------------------------------------------
function [coeffs] = lagrange_coeff(order,mu)
%   Calculates the coefficients of an interpolator filter with
%   a given interpolation order and a time shift 'mu'.
%   mu is centered around a symmetrical interval [-1,1]
mu = (mu-0.5)*2/order;

% coefficients are calculated following Lagrange interpolation
if (order == 1)      % linear
   coeffs = [0.5+0.5.*mu 0.5-0.5.*mu];
elseif (order == 2)     % quadratic
   coeffs = [      0.5.*mu + 0.5.*mu.^2 ...
              1            -  mu.^2 ...
                 -0.5.*mu + 0.5.*mu.^2 ];
elseif (order == 3)    % cubic
   mu2 = mu.*mu; % mu squared
   mu3 = mu2.*mu;% mu cubed
   coeffs = [ -0.0625 - 0.0625.*mu + 0.5625.*mu2 + 0.5625.*mu3 ...
```

MATLAB MODELS OF NON-IDEALITIES

```
              0.5625 + 1.6875.*mu - 0.5625.*mu2 - 1.6875.*mu3 ...
              0.5625 - 1.6875.*mu - 0.5625.*mu2 + 1.6875.*mu3 ...
             -0.0625 + 0.0625.*mu + 0.5625.*mu2 - 0.5625.*mu3 ];
end;
```

C.1.8 Clipping and quantization

```
function out = RX_CNQ(in,ENOB,CL)
% =========================================================================
% Book title: Digital Front-End Compensation for Emerging Wireless Systems
% Authors:    Francois Horlin (ULB), Andre Bourdoux (IMEC)
% Editor:     John Wiley & Sons, Ltd
% Date:       2008
% =========================================================================
%
% DESCRIPTION
%   This function clips and quantizes the signal stream(s). Each column
%   contains one stream. The number of bits of the quantizer is defined by
%   NB. The clip level is defined by CL. Each stream will be clipped at +/-
%   CL x RMS level of the stream
%
% INPUTS
%   - in        [length x Ns]: input signal where each column is a signal
%                              from a different antenna.
%   - ENOB              [1]: effective number of bits
%                              (can be non-integer)
%   - CL                [1]: clip level
%
% OUTPUTS
%   - out       [length x Ns]: Ns streams of output signals
[M N] = size(in);
sigmar = sqrt(mean(abs(real(in)).^2,1));
sigmai = sqrt(mean(abs(imag(in)).^2,1));
sigma_avg = (sigmar+sigmai)/2;
% Clip and quantize the real and imaginary parts separately
for k=1:N
    out(:,k) = cnq_enob(real(in(:,k)),ENOB,CL*sigma_avg(k)) +...
        i*cnq_enob(imag(in(:,k)),ENOB,CL*sigma_avg(k));
end
% ---------------------------------------------------
function y = cnq_enob(x,enob,cl);
% This function clips elements in vector x at the clipping level cl
% and quantizes them using ENOB bits. ENOB needs not be an integer
aux = round(2^enob)-1;
if cl~=Inf
    tmp = x/cl; % adjust to clipping level
    if enob==Inf
        tmp(find(tmp>1)) = 1;
        tmp(find(tmp<-1)) = -1;
    else
        if mod(aux,2)==0      % test if # of levels is even or uneven
            tmp = tmp.*aux/2;
            tmp(find(tmp>aux/2)) = aux/2;
            tmp(find(tmp<-aux/2)) = -aux/2;
            tmp = round(tmp);
            tmp = tmp/(aux/2);
            y = cl*tmp;
        else
```

```
                tmp = tmp.*aux/2;
                tmp(find(tmp>aux/2)) = aux/2;
                tmp(find(tmp<-aux/2)) = -aux/2;
                tmp = round(tmp+0.5)-0.5;
                tmp = tmp/(aux/2);
                y = cl*tmp;
        end
    end
else
    y = x;
end;
```

C.2   Transmitter non-idealities

C.2.1   Global TX non-idealities

```
% ==========================================================================
% Book title: Digital Front-End Compensation for Emerging Wireless Systems
% Authors:    François Horlin (ULB), André Bourdoux (IMEC)
% Editor:     John Wiley & Sons, Ltd
% Date:       2008
% ==========================================================================
%
% DESCRIPTION
%   This script provides an example to use the front-end TX non-ideality
%   functions.
%
% INPUTS     none
%
% OUTPUTS
%    - in     ideal input signal
%    - out    output signals affected by TX non-idealities
%    - a      structure containing all non-ideality parameters
%                  CNQ: [1x1 struct]
%                  SCO: [1x1 struct]
%                   IQ: [1x1 struct]
%                   PN: [1x1 struct]
%                  CFO: [1x1 struct]
%                   PA: [1x1 struct]

clear
% This generates a hypothetical transmit signal
Ns = 10000;
in = 5*exp(j*2*pi*(1:Ns)/128).';     % in must be a column vector

% We create variable "a", which is a structure containing
% all non-ideality parameters

% Use thse flags to enable/disable the corresponding functions
a.CNQ.on  = 1;
a.SCO.on  = 1;
a.IQ.on   = 1;
a.PN.on   = 1;
a.CFO.on  = 1;
a.PA.on   = 1;
```

MATLAB MODELS OF NON-IDEALITIES

```matlab
% Parameter settings
a.CNQ.enob   = 6;       % effective number of bits
a.CNQ.CL     = 2.5;     % clip level

a.SCO.sco    = 40e-6;   % value of SCO

a.IQ.eps     = 0.03;    % amplitude imbalance
a.IQ.dphi    = 0.03;    % phase imbalance
a.IQ.random  = 0;       % 0==fixed, 1==random

a.PN.dBc     = -20;     % integrated phase noise in dBc
a.PN.cutoff  = 100e3;   % cutoff frequency of PSD
a.PN.floor   = -120;    % phase noise floor in dBc/Hz
a.PN.fs      = 80e6;    % sampling frequency

a.CFO.fs     = 80e6;    % sampling frequency
a.CFO.cfo    = 40e-6;   % value of CFO

a.PA.backoff_dB = 2;    % back-off w.r.t. 1dB compression point

% Apply all TX non-idealities in order
out = in;  % This is needed in case no non-idealities are applied

% *** CLIP AND QUANTIZE ***
if(a.CNQ.on == 1)
    out = RX_CNQ(out,a.CNQ.enob,a.CNQ.CL);
end

% *** Sampling clock offset (SCO) ***
if (a.SCO.on == 1)
    out = RX_SCO(out,a.SCO.sco);
end

% *** I/Q mismatches ***
if (a.IQ.on == 1)
    out = TX_IQ(out,a.IQ.eps,a.IQ.dphi,a.IQ.random);
end

% *** ADD PHASE NOISE (PN) ***
if (a.PN.on == 1)
    out = RX_PN(out,a.PN.dBc,a.PN.cutoff,a.PN.floor,a.PN.fs);
end

% *** Carrier frequency offset (CFO) ***
if (a.CFO.on == 1)
    out = RX_CFO(out,a.CFO.fs,a.CFO.cfo);
end

% *** Non linear Amplification ***
if(a.PA.on == 1)
    out = TX_PA(out,a.PA.backoff_dB);
end
```

C.2.2 Clipping and quantization

This function is identical to the receiver clipping and quantization function (C.1.8).

C.2.3 Transmit IQ imbalance

```
function out = TX_IQ(in,eps,dphi,random);
% =============================================================================
% Book title: Digital Front-End Compensation for Emerging Wireless Systems
% Authors:    François Horlin (ULB), André Bourdoux (IMEC)
% Editor:     John Wiley & Sons, Ltd
% Date:       2008
% =============================================================================
%
% DESCRIPTION
%   This function applies TX IQ imbalance to the signal stream(s). Each column
%   contains one stream. If only one EPS or DPHI value is given, it is
%   applied to all streams. Otherwise a vector of EPS values and a vector of
%   DPHI values must be provided as input, each having a length equal to the
%   number of streams. The RANDOM parameter determines if EPS and DPHI are
%   used as such used or if random numbers are generated. In this latter
%   case, the amplitude is gausssian N(0,EPS^2) and DPHI is uniformly
%   distributed in the interval [-DPHI,DPHI].
%
% INPUTS
%   - in            [length x Ns]: input signal where each column is a signal
%                                  from a different antenna.
%   - EPS    [1] or [1 x length]: amplitude imbalance
%   - DPHI   [1] or [1 x length]: phase imbalance
%   - random                [1]: 1 --> random value, 0 --> fixed value
%
% OUTPUTS
%   - out           [length x Ns]: Ns streams of output signals
[M N] = size(in);
[P,Q] = size(eps);
[R,S] = size(dphi);

if P~=1 & Q~=1, error('eps must be scalar or vector'),end
if P==1 & Q==1, eps=repmat(eps,1,N);,
elseif max(P,Q)==N, eps  = reshape(eps,1,N);
else   error('eps length must be equal to # streams')
end
if R~=1 & S~=1, error('dphi must be scalar or vector'),end
if R==1 & S==1, dphi=repmat(dphi,1,N);,
elseif max(R,S)==N, dphi = reshape(dphi,1,N);
else   error('dphi length must be equal to # streams')
end

if random
    eps  = eps.*randn(1,N);        % linear value (not in %)
    dphi = dphi*2.*(rand(1,N)-0.5);  % in radians
else
    eps  = eps.*ones(1,N);                   % linear value (not in %)
    dphi = dphi.*ones(1,N);                  % in radians
end
for aa=1:N
    alph = cos(dphi(aa))+j*sin(dphi(aa))*eps(aa);
    beta = eps(aa)*cos(dphi(aa))+j*sin(dphi(aa));
    out(:,aa) = alph*in(:,aa)+beta*conj(in(:,aa));
end
```

C.2.4 Phase noise

This function is identical to the receiver phase noise function (C.1.4).

C.2.5 Carrier frequency offset

This function is identical to the receiver carrier frequency offset function (C.1.3).

C.2.6 Sampling clock offset

This function is identical to the receiver sampling clock offset function (C.1.7).

C.2.7 Nonlinear power amplifier

```
function out = TX_PA(in, Backoff_dB);
% =======================================================================
% Book title: Digital Front-End Compensation for Emerging Wireless Systems
% Authors:    Francois Horlin (ULB), Andre Bourdoux (IMEC)
% Editor:     John Wiley & Sons, Ltd
% Date:       2008
% =======================================================================
%
% DESCRIPTION
%   This function applies a third-order nonlinearity to the signal stream(s).
%   Each column contains one stream. The nonlinearity is determined by the
%   back-off wrt the 1 dB compression point.
%
% INPUTS
%   - in       [length x Ns]: input signal where each column is a signal
%                             from a different antenna.
%   - Backoff_dB       [1]: back-off wrt 1 dB compression point
%                             < 0 --> average power above P1dB
%                             > 0 --> average power below P1dB
%
% OUTPUTS
%   - out      [length x Ns]: Ns streams of output signals
%
% Transfer function of an nonlinear amplifier defined by its transfer
% function in amplitude y = alpha1 * x + alpha3 * x^3
%
% More explanations
% -----------------
% The amplifier is assumed to have
%   - unity gain
%   - 1 dB compression point normalized to 1
%   - transfer function in amplitude y = alpha1 * x + alpha3 * x^3
%
% The function will compute the RMS value of the signal and adapt its level
% according to the back-off requirement. Then the nonlinearity is
% introduced. At the end of the function, the RMS level is set back to its
% original value so that the energy of the input and output waveforms are
% identical. This is useful for BER simulations
%
% The formulas for the INPUT IP1, IP3 and saturation levels are:
%     1 dB compression point
%            IA1   = sqrt((1-10^(-1/20))*4*alpha1/(3*abs(alpha3)));
```

```
%           OA1     = IA1*alpha1*(10^(-1/20));
%
%      3rd order intercept point
%           IA3     = sqrt(4*alpha1/(3*abs(alpha3)));
%           OA3     = IA3*alpha1;
%
%      Saturation level (point where the tangent is horizontal
%           IAsat   = 2/3*sqrt(alpha1/abs(alpha3));
%           OAsat   = 4/9*alpha1*sqrt(alpha1/abs(alpha3));

[M N] = size(in);
for k=1:N
    % Parameters of nonlinear model
    alpha1 = 1;
    alpha3 = -(1-10^(-1/20))*4/3;   % This is an inversion of the formula
    % giving the input 1 dB compression point

    % RMS amplitude of signal
    Arms = sqrt(var(in(:,k)));

    % Brings signal to back-off dB below 1dB compression point
    in(:,k) = in(:,k)*10^(-Backoff_dB/20)/Arms;

    % Computes saturation values
    IAsat = 2/3*sqrt(alpha1/abs(alpha3));
    OAsat = 4/9*alpha1*sqrt(alpha1/abs(alpha3));

    % For the lowpass equivalent, multiply alpha3 by 3/4 !
    out(:,k)      = alpha1*in(:,k) + 0.75*alpha3*in(:,k).*abs(in(:,k)).^2;

    % Clipping to ensure monotonicity
    ind      = find(abs(in(:,k))>IAsat);
    out(ind,k) = OAsat*sign(in(ind,k));

    % Brings signal energy back to its original value
    YRMS = sqrt(var(out(:,k)));
    out(:,k) = out(:,k)/YRMS*Arms;
end
```

D

Mathematical Conventions

- Time-domain versus frequency-domain signals:

  $x(t)$ time-domain signal
  $\tilde{x}(f)$ frequency-domain signal

- Continuous versus discrete-time signals:

  $x(t)$ time-domain signal
  $x[n]$ time-domain sequence

- Statistical signals:

  $\mathcal{E}[\,.\,]$ expectation
  m_x mean
  σ_x^2 variance
  $R_{xy}(\tau)$ cross-correlation

- Complex signals:

  $|\,.\,|$ modulus
  $\angle(.)$ angle
  $\Re[\,.\,]$ real part
  $\Im[\,.\,]$ imaginary part

- Vector and matrices:

  $\underline{v}$ vector
  $\underline{\underline{m}}$ matrix
  $\underline{\underline{m}}_N$ square matrix of size N
  $\underline{\underline{m}}_{M \times N}$ matrix of size $M \times N$
  $[\underline{\underline{m}}]_{ij}$ element of $\underline{\underline{m}}$ located at row i and column j
  $\underline{\underline{0}}_{M \times N}$ matrix of zeros
  $\underline{\underline{I}}_N$ identity matrix
  $\underline{\underline{F}}_Q$ Fourier matrix

- Operators:

$\max[x(t)]$	maximum
$\log[x(t)]$	logarithm
$x(t) \star y(t)$	convolution
$(.)^T$	matrix transpose
$(.)^*$	complex conjugate
$(.)^H$	matrix complex conjugate transpose
$(.)^{-1}$	matrix inverse
$\underline{\underline{m}} \cdot \underline{\underline{n}}$	matrix product
$\underline{\underline{m}} + \underline{\underline{n}}$	matrix addition
$\underline{\underline{m}} - \underline{\underline{n}}$	matrix subtraction
$\underline{\underline{m}} \otimes \underline{\underline{n}}$	Kronecker product
$\underline{\underline{m}} \odot \underline{\underline{n}}$	Hadamard product
$\text{Tr}[\underline{\underline{m}}]$	trace of $\underline{\underline{m}}$
$\text{diag}[\underline{\underline{m}}]$	diagonal matrix of $\underline{\underline{m}}$

E
Abbreviations

3G	Third Generation
3GPP	Third Generation Partnership Project
3GPP LTE	Third Generation Partnership Project long term evolution
4G	Fourth Generation
16QAM	16 point quadrature amplitude modulation
64QAM	64 point quadrature amplitude modulation
AC	auto-correlation
ADC	analog-to-digital converter
AGC	automatic gain control
AM	amplitude modulation
AWGN	additive white Gaussian noise
BER	bit error rate
BLAST	Bell Laboratories layered space time
CC	convolutional code
CCK	complementary code keying
CDMA	code-division multiple access
CE	channel estimation
CFO	carrier frequency offset
CIR	channel impulse response
CMAC	complex multiply-and-accumulate
CPM	continuous phase modulation
CP	cyclic prefix
CP-CDMA	cyclic prefix code-division multiple access
CSD	cyclic shift diversity
CSI	channel state information
DA	data aided
DAB	digital audio broadcasting
DAC	digital-to-analog converter
DC	direct current
DD	decision directed

DF	decision feedback
DQPSK	differential quaternary phase shift keying
DFT	discrete Fourier transform
DL	downlink
DS	direct-sequence
DS-CDMA	direct-sequence code-division multiple access
DSL	digital subscriber line
DSSS	direct-sequence spread spectrum
DVB	digital video broadcasting
E-UTRA	evolved universal terrestrial radio access
EM	expectation maximization
ENOB	effective number of bits
ERBW	effective resolution bandwidth
FA	false alarm
FDD	frequency-division duplexing
FDE	frequency domain equalization
FDMA	frequency-division multiple access
FE	front-end
FEC	forward error code
FFT	fast Fourier transform
FHSS	frequency hopping spread spectrum
Gbps	gigabits per second
GFSK	Gaussian frequency shift keying
GI	guard interval
HSDPA	high speed downlink packet access
HSUPA	high speed uplink packet access
HT	high throughput
I	In-phase
IBI	inter-block interference
ICI	inter-carrier interference
IDFT	inverse discrete Fourier transform
IEEE	international electrical and electronical engineering
IF	intermediate frequency
IFFT	inverse fast Fourier transform
IMT	international mobile telecommunications
ISI	inter-symbol interference
ISM	industrial scientific and medical
LAN	local area network
LNA	low noise amplifier
LO	local oscillator
LPF	lowpass filter
LSB	least significant bit
LTE	long-term evolution
LTF	long training field
Mbps	megabits per second
MC	multi-carrier

MC-CDMA	multi-carrier code-division multiple access	
MCS	modulation coding schemes	
MIMO	multiple-input multiple-output	
MISO	multiple-input single-output	
ML	maximum likelihood	
MMSE	minimum mean square error	
MRC	maximum ratio combining	
MSE	mean square error	
MUI	multi-user interference	
OFDM	orthogonal frequency-division multiplexing	
OFDMA	orthogonal frequency-division multiple access	
P/S	parallel-to-serial	
PA	power amplifier	
PAPR	peak-to-average power ratio	
PDF	probability density function	
PLL	phase-locked loop	
PM	phase modulation	
PN	phase noise	
PSD	power spectral density	
PSK	phase shift keying	
Q	quadrature	
QAM	quadrature amplitude modulation	
QoS	quality-of-service	
QPSK	quaternary phase shift keying	
RF	radio frequency	
RMS	root mean square	
RX	receive	
S/P	serial-to-parallel	
SC	single-carrier	
SC-FDMA	single-carrier frequency-division multiple access	
SC-FDE	single-carrier frequency domain equalization	
SCBT	single-carrier block transmission	
SCO	sample clock offset	
SDM	space-division multiplexing	
SDMA	space-division multiple access	
SIC	successive interference canceller	
SIG	signal field	
SIMO	single-input multiple-output	
SIR	signal-to-interference ratio	
SISO	single-input single-output	
SNDR	signal-to-noise-and-distortion ratio	
SNIR	signal-to-noise-and-interference ratio	
SNR	signal-to-noise ratio	
SoA	state-of-the-art	
STBC	space–time block coding	
STC	space–time coding	

STF	short training field
SVD	singular value decomposition
TDD	time-division duplexing
TDMA	time-division multiple access
TR	technical report
TG	task group
TX	transmit
UL	uplink
UTRA	universal terrestrial radio access
V-BLAST	vertical Bell Laboratories layered space–time
VCO	voltage-controlled oscillator
VHT	very high throughput
WLAN	wireless local area network
WT	wireless terminal
XC	cross-correlation
ZF	zero forcing

Index

1-dB compression point, 47
3GPP long-term evolution, 28, 30, 205

A/D converter, 64, 104
Ad hoc non-linear model, 50
ADC sampling clock sharing, 44, 203
Additive white Gaussian noise, see AWGN
AGC, see Automatic gain control
AGC setting, 144
Alamouti scheme, see STBC
AM-to-AM, 50
AM-to-PM, 50
Amplifier, 44
Analog quadrature demodulation, 40, 43, 44
Analytical non-linear model, 47
Antenna mismatch, 122
Auto-correlation, 143, 152, 158
Automatic gain control, 138, 144
AWGN, 53

Back-off, 110, 112, 113, 116
Bandpass sampling, 40
Bandpass signal, 38
Burst detection, 138, 140

Calibration for reciprocity, 203
Carrier frequency offset, see CFO
Cascaded noise figure, 53
CDMA, 23
CFO, 45, 55, 73, 77, 89, 90, 117, 121, 128, 131, 153, 221
CFO acquisition, 138, 153, 194
CFO tracking, 138, 174, 175, 201
Channel bonding, 188
Channel estimation, 168, 196, 212
Channel sounding, 122
Channel state information at the transmitter, see CSIT
Channel tracking, 212
Clip power of OFDM, 105
Clip probability of OFDM, 104
Clipping, 46, 64, 103, 113
Closed-loop MIMO, 19, 22, 125
Coarse CFO acquisition, 138, 149, 194–196
Coarse timing acquisition, 138, 146, 194, 195

Code-division multiple access, see CDMA
Common phase, 77, 80, 90, 99
CP-CDMA, 25, 131
Cross-correlation, 152
CSIT, 19, 22, 122
Cyclic prefix, 9, 10
Cyclic-prefix code-division multiple access, see CP-CDMA

D/A converter, 64
DC offset, 45, 63
DFT-spread OFDM, 13, 30
Digital quadrature demodulation, 41
Direct conversion receiver, 43
Direct-sequence code-division multiple access, see DS-CDMA
Distributed transmission, 28, 30, 207
Doppler shift, 55
Down conversion, 39, 45
Downlink, 24, 29, 128
DS-CDMA, 23

Effective number of bits, see ENOB
Effective resolution bandwidth, 65
EM, 159
Energy detection, 141
ENOB, 65
Equalization, 9, 13
Expectation-maximization, 159

Fast Fourier Transform, see FFT
FDMA, 28
FFT, 9
Fine CFO acquisition, 138, 151, 153, 194, 196
Fine timing acquisition, 138, 150, 194, 196
Frame, 190, 208
Frequency domain equalization, 9, 13
Frequency-dependent IQ imbalance, 60, 86, 168, 172
Frequency-division multiple access, see FDMA
Front-end, 37
Front-end architecture, 38
Front-end model, 73

Greenfield, 190

High-speed dowlink packet access, see HSDPA
High-speed uplink packet access, see HSUPA
HSDPA, 23, 205
HSUPA, 23, 205

IBI, 11
ICI, 77, 80, 82, 99
Ideal receiver, 38
Ideal transmitter, 38
IEEE802.11, 9, 186
IEEE802.11-VHT, 186
IEEE802.11a, 9, 13, 186
IEEE802.11b, 186
IEEE802.11g, 9, 13, 186
IEEE802.11n, 15, 186, 188
IEEE802.16e, 28
Image frequency, 45
Inter-block interference, see IBI
Inter-carrier interference, see ICI
Inter-symbol interference, see ISI
Interpolation, 69
IQ imbalance, 45, 58, 73, 82, 89, 93, 128, 131, 153, 168
IQ imbalance compensation, 153, 168
IQ imbalance estimation, 153, 196
ISI, 90, 91, 93, 99
Iterative acquisition, 159

Joint acquisition, 153
Joint model CFO, SCO and IQ, 73, 89, 154

Likelihood function, 157
Linearly pre-coded OFDM, 9, 13
Local oscillator sharing, 44, 203
Localized transmission, 28, 30, 207
Long-term evolution, see 3GPP long-term evolution
Low-IF receiver, 44
Low-pass signal, 38

Matlab files, 47, 231
Maximum likelihood, 157, 229
Maximum ratio combining, see MRC
MC-CDMA, 24, 128
MIMO, 15, 117
MIMO front-end architecture, 44
MIMO-OFDM, 15
Minimum mean square error, 9, 18, 23, 25, 227
Mixer, 45
ML, 157, 229
ML channel estimator, 170, 216, 229
MMSE linear detector, 9, 18, 23, 25, 227
MMSE precoder, 22
Modified Rapp model, 50
MRC, 17, 120
MUI, 128, 131
Multi-carrier code-division multiple access, see MC-CDMA
Multi-input multi-output, see MIMO

Multi-user, 21
Multi-user interference, see MUI
Multi-user joint detection, 25
Multiple antennas, 15
Multiple users, 126
Multiplicative noise, 58

Noise figure, 53
Non-reciprocity, 122, 125, 203
Non-reciprocity compensation, 203
Nonlinear amplifier, 110
Nonlinear model, 47
Nonlinearity, 47, 103
Nyquist sampling, 41

OFDM, 9, 72, 98, 135
OFDMA, 28, 128, 207
Ordered SIC, 18, 23
Orthogonal frequency-division multiple access, see OFDMA
Orthogonal frequency-division multiplexing, see OFDM
Oversampling, 64

PAPR, 53, 136
Peak-to-average power ratio, see PAPR
Phase noise, 45, 56, 98
Phase noise tracking, 174, 179
Phase-locked loop, 56, 176
Pilot symbols, 137, 208
PLL, 56, 176
PN, 56, 98
Power amplifier, 110
Power spectral density, 51, 108, 111, 113
Pre-coder, 9, 13
Preamble, 136, 154, 190, 208

Quantization, 46, 64, 106, 113
Quantization noise, 64, 106
Quantization variance, 64, 106

Rapp model, 50
Receiver IQ Imbalance, 59, 73, 82, 89, 93, 128, 131, 153, 168
Reciprocity, 19, 22
Resistor noise, 53

Saleh model, 50
Sample clock offset, see SCO
Sampling jitter, 46, 67
Saturated output power, 47
Saturation, 47
SC-FDE, 13, 89, 98
SC-FDMA, 30, 131, 207
SCO, 46, 65, 73, 80, 89, 91, 117, 223
SCO estimation, 138, 194
SCO tracking, 138, 174, 179, 201
SDM, 18
SDMA, 21

INDEX

SIC, 18, 23
Single-carrier, *see* SC-FDE
Single-carrier frequency domain equalization, *see* SC-FDE
Single-carrier frequency-division multiple access, *see* SC-FDMA
Software defined radio, 4
Space–time block coding, *see* STBC
Space–time coding, *see* STC
Space-division multiple access, *see* SDMA
Space-division multiplexing, *see* SDM
Spatial streams, 18
Spectral regrowth, 51, 108, 111, 113, 116
STBC, 17, 120
STC, 17
Sub-sampling, 41
Successive interference canceller, *see* SIC
Super-heterodyne receiver, 40

Technology scaling, 4
Thermal noise, 53

Third-order intercept point, 47
Timing acquisition, 138, 150, 194
Tracking loop, 174, 179, 201
Transceiver architecture, 38
Transmitter and receiver IQ imbalance, 62
Transmitter IQ imbalance, 60, 131

Up conversion, 39, 45
Uplink, 25, 30, 131, 210
Uplink synchronization, 210

Very high throughput, 186

Wireless local area network, *see* WLAN
Wireless system, 2
WLAN, 13, 135, 185

Zero forcing, 9, 22, 23
Zero-IF receiver, 43

Printed in the United States
By Bookmasters